Network Optimization

V. K. Balakrishnan

Department of Mathematics
University of Maine
Orono, Maine
USA

CHAPMAN & HALL

London · Glasgow · Weinheim · New York · Tokyo · Melbourne · Madras

वागर्थाविव संपृक्तौ वागर्थप्रतिपत्तये।
जगतः पितरौ वन्दे पार्वतीपरमेश्वरौ॥

Opening verse in Kalidasa's Raghuvamsam

(Salutations to Parvathi and Parameswara – the Mother and the Father of the Universe – who are connected to each other the way a word and its meaning are linked for the complete understanding of the meaning of a word.)

Dedicated to the fond memory of
my parents
who convinced me that a quest for knowledge
is the most worthwhile of all our pursuits

Published by Chapman & Hall, 2–6 Boundary Row, London SE1 8HN, UK

Chapman & Hall, 2–6 Boundary Row, London SE1 8HN, UK

Blackie Academic & Professional, Wester Cleddens Road, Bishopbriggs, Glasgow G64 2NZ, UK

Chapman & Hall GmbH, Pappelallee 3, 69469 Weinheim, Germany

Chapman & Hall USA, One Penn Plaza, 41st Floor, New York NY 10119, USA

Chapman & Hall Japan, ITP-Japan, Kyowa Building, 3F, 2-2-1 Hirakawacho, Chiyoda-ku, Tokyo 102, Japan

Chapman & Hall Australia, Thomas Nelson Australia, 102 Dodds Street, South Melbourne, Victoria 3205, Australia

Chapman & Hall India, R. Seshadri, 32 Second Main Road, CIT East, Madras 600 035, India

First edition 1995

© 1995 V. K. Balakrishnan

Typeset in 10/12pt Times by Thomson Press (India) Ltd., New Delhi

Printed and bound in Great Britain by Hartnolls Limited, Bodmin, Cornwall

ISBN 0 412 55670 7

A catalogue record for this book is available from the British Library

27/2/96.

Contents

Preface

Problems in network optimization arise in all areas of technology and industrial management. The topic of network flows, in particular, has applications in such diverse fields as chemistry, communications, computer science, economics, engineering, management science, scheduling and trans- portation, to name but a few. The aim of this book is to present the important concepts of network optimization in a concise textbook suitable for upper- level undergraduate students in computer science, mathematics and opera- tions research. While discussing algorithms, the emphasis in this book is more on clarity and plausibility than on complexity considerations. At the same time, rigorous arguments have been used in proving the basic theorems. A course in introductory linear algebra is the only prerequisite needed to follow the material presented in the text.

The basic graph theory concepts are presented in Chapter 1, along with problems related to spanning trees and branching. Network flow problems are then treated in detail in Chapters 2–4. It is the network simplex algorithm that holds these three chapters together. The theory of matching in both bipartite and nonbipartite graphs is covered in Chapter 5. The relation between network flow problems and certain deep theorems in combinatorics – a topic usually not covered in undergraduate texts – is established in detail in sections 2.6 and 4.8, culminating in a grand equivalence theorem as displayed in Figure 4.10. Each chapter ends with several exercises, ranging from routine numerical ones to challenging theoretical problems, and a student interested in learning the subject is expected to solve as many of these as possible.

This book is based on topics selected from courses on graph theory, linear programming, discrete mathematics, combinatorics and combinatorial optimization taught by me to undergraduate and beginning graduate students in computer science, electrical engineering, forest management and mathematics at the University of Maine during the last decade. The contribution of my students is implicit on every page of this text. It is indeed a pleasure to acknowledge their participation and excitement in this collective learning process.

My indebtedness in writing this monograph also encompasses many sources, including the tomes mentioned in the reading list and in the list of references at the end of the book and several mathematicians with whom I have come into contact in the last few years at national and international conferences and workshops. In particular, I would like to thank Vašek Chvátal, Victor Klee, George Nemhauser, James Orlin, Christos Papadimitriou, Robert Tarjan and Herbert Wilf – creative mathematicians who also happen to be truly inspiring teachers. Eugene Lawler (who is no longer with us) was always a source of inspiration and guidance for the completion of this project, and I am beholden to him for ever. He will be dearly missed. Special thanks also go to the reviewer from Chapman & Hall who patiently read the manuscript more than once and offered many constructive suggestions and helpful comments. If there are still errors or omissions, the responsibility is entirely mine.

I am grateful to my colleagues and the administrators at the University of Maine for providing me with necessary facilities and opportunities to undertake and finalize this project.

In conclusion, I would like to express my profound gratitude to the editorial and production staff at Chapman & Hall, particularly to my commissioning editor, Achi Dosanjh, for her unfailing cooperation and continual support during the entire reviewing and editing process of this book, and to my sub-editor, Howard Davies, for taking care of several last-minute corrections along with his multifarious sub-editing responsibilities. I also thank with great pleasure my initial commissioning editor, Nicki Dennis, for the strong encouragement she gave me while accepting my proposal. Finally, I would like to add that I owe a great deal to the keen interest and abiding affection my family has shown me at every stage of this work.

V. K. Balakrishnan
University of Maine
January 1995

1

Trees, arborescences and branchings

1.1 SOME GRAPH THEORY CONCEPTS

Graphs and digraphs

A **graph** $G = (V, E)$ consists of a finite nonempty set V and a collection E of unordered pairs from V (i.e. two-element subsets of V). Every element in V is called a **vertex** of the graph, and each unordered pair in E is called an **edge** of the graph. The edge $e = \{x, y\}$ is an edge between the two vertices x and y which is **incident** to both x and y. Two vertices are **adjacent** to each other if there is an edge between them: the edge in that case is said to **join** the two vertices. Two or more edges that join the same pair of vertices are called **parallel** edges. A graph without any parallel edges is a **simple graph**, in which case the collection E is a set. Otherwise G is **multigraph**. A graph G' is a **subgraph** of the graph G if every vertex of G' is a vertex of G and every edge of G' is an edge of G.

A **directed graph** or **digraph** consists of a finite set V of vertices and a collection A of ordered pairs from V called the **arcs** of the digraph. The digraph is a **simple digraph** if A is a set. If $a = (x, y)$ is an arc, then a is an arc from the vertex x to the vertex y and is incident (adjacent) from x and incident (adjacent) to y. Two vertices are **nonadjacent** if there is no arc from one to the other. The **underlying graph** G of a digraph D is the graph G obtained from D by replacing each arc (x, y) by an edge $\{x, y\}$.

In this book, unless otherwise mentioned, all graphs are simple graphs and all digraphs are simple digraphs.

If we associate one or more real numbers with each edge (or arc) of a graph (or a digraph), the resulting structure is known as a **weighted graph** or a **network**.

Connectivity

A **path** between vertex x_1 and vertex x_r in a graph is a sequence $x_1, e_1, x_2, e_2, x_3, e_3, \ldots, e_{r-1}, x_r$ where $x_1, x_2, x_3, \ldots$ are vertices and e_k is the edge between

x_{k-1} and x_k for $k = 1, 2, \ldots, r$. This path is represented as $\langle x_1, x_2, x_3, \ldots, x_r \rangle$ or as $x_1 - x_2 - x_3 - \cdots - x_r$ without mentioning the edges explicitly. If $x_1 = x_r$ it is a **closed path**. It is a **simple path** between x_1 and x_r if $x_1, x_2, \ldots, x_r$ are distinct. A **circuit** in a graph is a closed path in which the edges are all distinct. A **cycle** is a circuit in which the vertices are all distinct. Obviously every cycle is a circuit. A cycle with k vertices ($k \geqslant 3$) has k edges and is called an **even cycle** if k is even. Otherwise it is an **odd** cycle.

Two vertices x and y in a graph are **connected** to each other if there is a path between them. A graph is a **connected graph** if there is a path between every pair of vertices in it. Otherwise it is a **disconnected** graph. A connected subgraph H of the graph G is called a **component** of G if there is no connected subgraph H' (other than H) such that H is a subgraph of H'.

If $G = (V, E)$ and $E = E' \cup \{e\}$, then $G' = (V, E')$ is the subgraph of G obtained from G by **deleting** the edge e from G. An edge e in a connected graph G is a **bridge** if its deletion from G yields a disconnected subgraph.

A **directed path** from a vertex x_1 to a vertex x_r in a digraph is a sequence $x_1, a_1, x_2, a_2, \ldots, a_{r-1}, x_r$ in which the x_i are vertices and a_i is the arc from x_i to x_{i+1}, where $i = 1, 2, \ldots, r-1$. This path may be represented as $x_1 \to x_2 \to \cdots \to x_r$. This is a closed directed path if $x_1 = x_r$. A **simple directed path** from x to y is a directed path in which all the vertices are distinct. A **directed circuit** is a closed directed path with distinct arcs, and a **directed cycle** is a closed directed path with distinct vertices. Two vertices x and y in a digraph form a **strongly connected pair** if there is a directed path from x to y and a directed path from y to x. A digraph is a **strongly connected digraph** if every pair of vertices in the digraph is a strongly connected pair. A digraph is **weakly connected** if its underlying graph (multigraph) is connected.

A simple path in a graph is called a **Hamiltonian path** if it passes through every vertex of the graph. A closed Hamiltonian path is a **Hamiltonian cycle**. An **Eulerian circuit** in a graph is a circuit which contains every edge of the graph. The definitions in the case of digraphs are analogous.

Bipartite graphs and complete graphs

If the set V of vertices of a graph $G = (V, E)$ is partitioned into two subsets X and Y such that every edge in E is between some vertex in X and some vertex in Y, then the graph is called a **bipartite graph** and is denoted by $G = (X, Y, E)$.

Theorem 1.1
A graph with three or more vertices is bipartite if and only if it has no odd cycles.

Proof
If $G = (X, Y, E)$ is a bipartite graph and if C is a cycle in G, obviously it should have an even number of vertices since the vertices in C are alternately from X and Y.

On the other hand, suppose $G = (V, E)$ has no odd cycles. Observe that a graph G is bipartite if and only if every component of G is bipartite. So we assume without loss of generality that G is connected. The number of edges in a simple path P between two vertices u and v is the length of P. A path P of minimum length between u and v is a shortest path and the number of edges in such a path is denoted $d(u, v)$.

Let u be any vertex in G. Define $X = \{x \in V : d(u, x)$ is even$\}$ and $Y = V - X$. We now show that whenever v and w are any two vertices in X (or in Y), there is no edge in E joining v and w.

Case 1. Let v be any vertex in X other than u. Since there is a path of even length between u and v, there cannot be an edge between u and v since there is no odd cycle in G.

Case 2. Let v and w be two vertices in X other than u. Assume there is an edge e joining v and w. Let P be a shortest path of length $2m$ between u and v and Q be a shortest path of length $2n$ between u and w. If P and Q have no common vertices other than u, then these two paths and the edge e will constitute an odd cycle. If P and Q have common vertices, let u' be the common vertex such that the subpath P' and u' and v and the subpath Q' between u' and w have no vertex in common. Since P and Q are shortest paths, the subpath of P between u and u' is a shortest u–u' path. The subpath of Q between u and u' is also a shortest u–u' path. So both these shortest u–u' paths are of equal length k. Thus the length of P' is $2m - k$ and the length of Q' is $2n - k$. In this case P', Q' and the edge e together constitute a cycle of length $(2m - k) + (2n - k) + 1$ which is an odd integer.

Case 3. Suppose v and w are in Y. Then the length of a shortest path P between u and v is odd. As in case 2, the subpath P' from u' to v of length $(2m - 1) - k$, the subpath Q' from u' to w of length $(2n - 1) - k$ and the edge will form an odd cycle.

Case 4. Suppose u is the only vertex in X. Then every edge is between u and some vertex in Y.

A graph $G = (V, E)$ is said to be **complete** if there is an edge between every pair of vertices in the graph. A bipartite graph $G = (V, W, E)$ is complete if there is an edge between every vertex v in V and every vertex w in W. A complete graph with n vertices is denoted by K_n. If $G = (V, W, E)$ is a complete bipartite graph with m vertices in V and n vertices in W, then G is denoted by $K_{m,n}$.

Degrees, indegrees and outdegrees

The number of edges incident at the vertex of a graph or a multigraph G is the **degree** of the vertex. A vertex of G is **odd** if its degree is odd. Otherwise it is an **even** vertex. In computing the degrees of the vertices, each edge is considered twice. Thus the sum of the degrees of all the vertices of G is equal to twice the number of edges in G and consequently the number of odd vertices of G is even.

The number of arcs incident to a vertex in a digraph is the **indegree** of the vertex, and its **outdegree** is the number of arcs incident from that vertex. The **degree** of a vertex in a digraph is the sum of its indegree and outdegree. In any digraph, the sum of the indegrees of all the vertices and the sum of the outdegrees of all the vertices are both equal to the total number of arcs in the digraph since each arc is incident from one vertex and incident to another vertex. As in the case of graphs, the sum of the degrees of all the vertices in a digraph is twice the number of arcs in it.

Incidence matrices and totally unimodular matrices

Suppose $G = (V, E)$ is a graph where $V = \{x_1, x_2, x_3, \ldots, x_m\}$ and $E = \{e_1, e_2, e_3, \ldots, e_n\}$. Then the **incidence matrix** of the graph G is the $m \times n$ matrix $A = [a_{ik}]$, where each row corresponds to a vertex and each column corresponds to an edge such that if e_k is an edge between x_i and x_j then $a_{ik} = a_{jk} = 1$ and all the other elements in column k are 0. The incidence matrix of a digraph $D = [V, E]$ where $V = \{v_1, v_2, \ldots, v_m\}$ and $E = \{a_1, a_2, \ldots, a_n\}$ is the $m \times n$ matrix $[a_{ik}]$ such that each row corresponds to a vertex and each column corresponds to an arc with the following property: if a_k is the arc from vertex v_i to vertex v_j, then $a_{ik} = -1$, $a_{jk} = 1$ and all other elements in column k which correspond to the arc a_k are 0.

A matrix A with integer entries is called a **totally unimodular** (TU) matrix if the determinant of every (square) submatrix B of A is 0 or 1 or -1. The determinant of a square matrix B is denoted as det B. Obviously in a TU matrix every element is either 0 or 1 or -1.

Notice that if A is a TU matrix and if B is any nonsingular submatrix of A, the unique solution of the linear system $BX = C$, whenever C is an integer vector, is also an integer vector since each component of the solution vector is of the form p/q, where p is the determinant of an integer matrix and therefore an integer and q is the determinant of B which is either 1 or -1. TU matrices play an important role in network and combinatorial optimization problems in which we seek solution vectors with integer components.

Theorem 1.2
The matrix $A = [a_{ij}]$, in which every element is 0 or 1 or -1, is totally unimodular if it satisfies the following two properties:

1. No column can have more than two nonzero elements.
2. It is possible to partition the set I of rows of A into sets I_1 and I_2 such that if a_{ij} and a_{kj} are the two nonzero elements in column j, then row i and row k belong to the same subset of the partition if and only if they are of opposite sign.

Proof

Let C be any $k \times k$ submatrix of $\mathbf{A}$. The proof is by induction on k. If $k = 1$, then the theorem is true. Suppose it is true for $k - 1$. We have to prove it is true for k. Let C' be any $(k - 1) \times (k - 1)$ submatrix of C. By induction hypothesis, det C' is 0 or 1 or -1. So we have to establish that the det C is 0 or 1 or -1.

There are three different possibilities: (i) C has a column such that all the elements in that column are zero and this implies det $C = 0$. (ii) C has a column with exactly one nonzero entry which could be either 1 or -1. Then expanding along this column, det C is either det C' or $-$ det C'. Thus det C is 0 or 1 or -1. (iii) Every column of C has exactly two elements.

Suppose $E = \{r_1, r_2, \ldots, r_k\}$ is the set of the rows of C. By hypothesis, this set of k rows is partitioned into two subsets I_1 and I_2. Without loss of generality, let us assume that I_1 is the set of the first p rows and I_2 is the set of the remaining $k - p$ rows. (It is possible that $p = 0$.) By the way these two sets are constructed, it is easy to see that $r_1 + r_2 + \cdots + r_p = r_{p+1} + r_{p+2} + \cdots + r_k$. So C is linearly dependent and thus det $C = 0$.

Corollary 1.3

A matrix in which each element is 0 or 1 or -1 is totally unimodular if in each column there is at most one $+1$ and at most one -1. In particular, the incidence matrices of digraphs and bipartite graphs are totally unimodular.

Theorem 1.4

The following properties are equivalent in a graph: (i) G is bipartite; (ii) G has no odd cycle; and (iii) the incidence matrix of G is totally unimodular.

Proof

The equivalence of (i) and (ii) has been already established by Theorem 1.1. If G is bipartite, its incidence matrix is a TU matrix by Corollary 1.3.

Suppose the incidence matrix of an arbitrary graph $G = (V, E)$ is a TU matrix. If the graph G is not bipartite, there should be at least one odd cycle in the graph. Let $V = \{1, 2, 3, \ldots, n\}$. Assume the first $2k + 1$ vertices in V constitute an odd cycle: $1 - 2 - 3 - \cdots - 2k + 1 - 1$, Let $\mathbf{A}$ be the incidence matrix of the graph in which the first $2k + 1$ rows correspond to the first $2k + 1$ vertices and the first $2k + 1$ columns correspond to the edges $\{1, 2\}$, $\{2, 3\}$, $\{3, 4\}, \ldots, \{2k, 2k + 1\}$, $\{2k + 1, 1\}$, respectively. Let C be a submatrix formed by the first $2k + 1$ rows and the first $2k + 1$ columns of the incidence matrix. Then the determinant of C is 2. (In fact, the determinant of the incidence matrix of an odd cycle is always 2 or -2.) This contradicts the fact that $\mathbf{A}$ is totally unimodular.

1.2 SPANNING TREES

Forests and trees

A graph in which no subgraph is a cycle is called an **acyclic graph** or a **forest**. A connected acyclic graph is a **tree**.

Theorem 1.5
A graph is a tree if and only if there is a unique simple path between every pair of vertices in the graph.

Proof
Let G be a graph such that between every pair of vertices there is a unique simple path. So G is connected. If G is not a tree, then there is at least one cycle in G creating two simple paths between every pair of vertices in this cycle. So G is a tree.

On the other hand, suppose G is a tree and x and y are two vertices in G. Since G is connected there is a simple path P between x and y. Suppose P' is another simple path between x and y. If the two paths are not the same, then let $e = \{v_i, v_{i+1}\}$ be the first edge in P that is not in P' as we go from x to y in the graph along the edges. Let W and W' be the set of intermediate vertices between v_i and y in P and P', respectively. If W and W' have no vertices in common, there is a cycle consisting of all the vertices in W and W' and the vertices v_i and y. If W and W' have common vertices, then let w be the first common vertex as we go from v_i to y along either P or P'. Then we have a cycle in G using the vertices in P between v_i and w and the vertices in P' between w and v_i. Thus in any case G has a cycle, which is a contradiction.

Theorem 1.6
A graph is a tree if and only if every edge in it is a bridge.

Proof
Suppose $e = \{x, y\}$ is any edge in a graph G. Then $\langle x, e, y \rangle$ is a path between x and y in G. If G is a tree, this is the only path between x and y, and if this path is deleted then the vertices x and y will not be connected. Thus in a tree every edge is a bridge.

On the other hand, suppose G is a graph in which every edge is a bridge. So G is connected. Suppose G is not a tree. Then there is at least one cycle C in G. Let x and y be adjacent vertices in C. Then the edge $e = \{x, y\}$ cannot be a bridge in G since there is another path between x and y using the other vertices of the cycle. This contradicts the hypothesis that every edge is a bridge. Thus G is a tree.

Theorem 1.7
(i) A connected graph with n vertices is a tree if and only if it has $n-1$ edges.
(ii) An acyclic graph with n vertices is a tree if and only if it has $n-1$ edges.

Proof
(i) Suppose G is a tree with n vertices. We prove by induction on n that G has $n-1$ edges.

This is true when $n = 1$. Suppose it is true for all m, where $1 < m < n$. If we delete an edge $e = \{x, y\}$ from G, we get two subgraphs $H = (V, E)$ and

$H' = (V', E')$, with k and k' vertices, respectively, such that there is no vertex common to V and V'. Since both k and k' are less than n, H has $k - 1$ edges and H' has $k' - 1$ edges. So both H and H' together have $k + k' - 2 = n - 2$ edges. If we combine H and H' by using the edge e, we get the graph G. So G has $(n - 2) + 1 = n - 1$ edges.

On the other hand, let $G = (V, E)$ be a connected graph with n vertices and $n - 1$ edges. Suppose G is not a tree. Then there is an edge in G which is not a bridge. If we delete this edge we have a connected subgraph $G' = (V, E')$. Continue this process till we get a connected subgraph $H = (V, F)$ in which every edge is a bridge. So H is a tree with n vertices. So H has $n - 1$ edges leading to a contradiction, since G has $n - 1$ edges.

(ii) If an acyclic graph with n vertices is a tree, then it has $n - 1$ edges by (i) above.

On the other hand, let $G = (V, E)$ be an acyclic graph with n vertices and $n - 1$ edges. Suppose G is not connected. Then there are r connected subgraphs $G_1, G_2, \ldots, G_r$, where $G_i = (V_i, E_i)$, the cardinality of V_i is n_i for each i and $\{V_1, V_2, \ldots, V_r\}$ is a partition of V. Each G_i is a connected component of G. Since G is acyclic, each G_i is acyclic and therefore a tree with $n_i - 1$ edges. Thus the total number of edges in G is $(n_1 - 1) + (n_2 - 1) + \cdots + (n_r - 1) = n - r$. But the number of edges in G is $n - 1$. Thus $r = 1$, which implies that G is connected and hence a tree.

Spanning trees

If $G = (V, E)$ is a graph and $T = (V, F)$ is a subgraph which is also a tree, then T is a **spanning tree** in the graph. An edge in E which is not in F is called a **chord** of T. The following two theorems are immediate consequences of Theorems 1.5–1.7.

Theorem 1.8
A graph is connected if and only if it has a spanning tree.

Theorem 1.9
Suppose $G = (V, E)$ is a simple graph with n vertices and $H = (V, E)$ is a subgraph. If H satisfies any two of the following three properties then it satisfies the third property also:

(i) H is connected.
(ii) H is acyclic.
(iii) H has $n - 1$ edges.

If a graph $G = (V, E)$ is connected, starting from any vertex it is possible to visit all the other vertices by searching these vertices. There are two ways of accomplishing this: the **breadth-first search** (BFS) and the **depth-first search** (DFS).

The BFS method proceeds as follows. Starting from any vertex v, visit all the vertices which are adjacent to v using the edges which join v and these adjacent vertices. Suppose these vertices are $v_1, v_2, \ldots, v_k$. Now start from v_1 and visit all the unvisited vertices which are adjacent to v_1 using the edges joining v_1 and these adjacent vertices. Suppose these vertices are $v_{11}, v_{12}, \ldots, v_{1k}$. Then we start from v_2 and continue this process. Proceed likewise till we visit all the vertices (as yet unvisited) adjacent to v_k. Now start from v_{11} and visit all the unvisited vertices adjacent to v_{11}. Then from v_{12} and so on. Continue this process till all vertices of G are visited. If F is the set of the edges used to visit all the vertices of G, then (V, F) is a spanning tree in G known as the **BFS spanning tree rooted at** v_1.

In the DFS method, start from any vertex v_0 and use an edge e_{01} to visit an adjacent vertex v_1. Then use an edge e_{12} to visit the vertex v_2 which is adjacent to v_1. In general, when we are at a vertex v_i, we use an edge v_{ij} to visit an adjacent vertex v_j if v_j has not been visited. If v_i has no unvisited adjacent vertex, we backtrack to v_{i-1} and explore further. Eventually this backtracking takes us back to the vertex v from which we started. If F is the set of the edges used in this procedure, then (V, F) is a spanning tree known as the **DFS spanning tree rooted at** v_0.

Disconnecting sets and cutsets

If $G = (V, E)$ is a connected graph, then any subset D of E is a **disconnecting set** if the deletion of all the edges in D from the graph makes it disconnected. A disconnecting set D is a **cutset** if no proper subset of D is a disconnecting set. Suppose the set V is partitioned into two subsets X and Y. If $D = (X, Y)$ is the set of all edges in E of the form $\{x, y\}$ where $x \in X$ and $y \in Y$, then D must be a disconnecting set. But D need not be a cutset.

When will a partition of the set of vertices of a connected graph define a cutset? We have the following theorem in this context.

Theorem 1.10
A disconnecting set $D = (X, Y)$ in a graph $G = (V, E)$ will be a cutset in G if $G' = (V, E - D)$ has exactly two connected components.

Proof
If D is not a cutset, then there is a proper subset D' of D which is a cutset. Suppose $e = \{x, y\}$, where $x \in X$ and $y \in Y$ is an edge in D but not in D'. If v is any vertex in X and w any vertex in Y, by hypothesis there is a path between v and x passing through vertices exclusively from X and there is a path between y and w passing through vertices exclusively from Y. Consequently the vertex v in X and the vertex w in Y are connected as long as the edge e remains undeleted. In other words, D' is not a disconnecting set and so it cannot be a cutset.

Corollary 1.11

If e is any edge of a spanning tree T in a connected graph G, the deletion of e from the tree defines a partition of the set of vertices of T (and therefore of G) into two subsets X and Y such that $D = (X, Y)$ is a cutset of the graph G.

Observe that even though an arbitrary partition of the set V of the connected graph $G = (V, E)$ into two subsets need not define a cutset, a cutset D in G always defines a partition of the set of vertices into two sets. Specifically, if the deletion of the edges of a cutset D from the graph creates two connected subgraphs $G' = (V', E')$ and $G'' = (V'', E'')$ then $\{V'', V''\}$ is a partition of V and D is the disconnecting set (V', V'').

Theorem 1.12

Suppose C is a cycle, D is a cutset and T is a spanning tree in a graph G. Then: (i) the number of edges common to C and D is even; (ii) at least one edge of C is a chord of T; and (iii) at least one edge of D is an edge of T.

Proof

(i) Let $D = (X, Y)$. If all the vertices of C are exclusively in X (or exclusively in Y) then C and D cannot have any edges in common. Suppose this is not the case. Then C has two vertices x and y where $x \in X$ and $y \in Y$. Then the cycle C which starts from x and ends at x will have to use the edges from D an even number of times.

(ii) A chord of T is by definition an edge of the graph G which is not an edge T. If no edge of C is a chord of T, then C is a subgraph of T. But C is a cycle. This is a contradiction.

(iii) If no edge of D is an edge of T, the deletion of all the edges belonging to D will not disconnect T. So T will continue as a spanning tree of G, implying that D is not a cutset.

If $e = \{x, y\}$ is a chord of T, the unique cycle in G consisting of the edge e and the edges of the unique path in T between x and y is called the **fundamental cycle** of G relative to T with respect to e and is denoted by $C^T(e)$, or by $C(e)$ if the tree T is fixed beforehand.

It follows from Corollary 1.11 that corresponding to each edge e of a spanning tree T of a connected graph G there is a unique cutset called the **fundamental cutset** of T with respect to the edge e which is denoted by $D^T(e)$, or by $D(e)$ if the tree is fixed beforehand.

Thus every spanning tree of a connected graph G with n vertices and m edges defines a set of $n - 1$ fu ndamental cutsets and a set of $m - n + 1$ fundamental cycles.

There are two theorems relating the concepts of fundamental cycles and fundamental cutsets of a spanning tree.

Theorem 1.13

Let e be an edge of a spanning tree T in a graph and f be any edge (other than e) in the fundamental cutset $D(e)$ defined by e. Then: (i) f is a chord of T and e is an edge of the fundamental cycle $C(f)$ defined by f; and (ii) e is not an edge of the fundamental cycle $C(e')$ defined by any chord e' which is not in $D(e)$.

Proof

(i) If f is not a chord of T, then it is an edge of the spanning tree. In that case T cannot be acyclic, which is a contradiction. So f is a chord of T defining a fundamental cycle $C(f)$. Now $D(e)$ is the union of $\{e, f\}$ and a set A of chords of T, and $C(f)$ is the union of $\{f\}$ and a set B of edges of T. The intersection of A and B is empty. By the previous theorem the intersection of $D(e)$ and $C(f)$ should have an even number of edges. So e is in B and therefore in $C(f)$.

(ii) Now $C(e')$ is the union of $\{e'\}$ and a set X of edges of T. The set $D(e)$ is the union of $\{e\}$ and the set A as in (i) above. Suppose e is in $C(e')$. This implies that e is in X. Since $C(e')$ and $D(e)$ have an even number of edges in common and since the intersection of A and X is empty, the edge e' is in $D(e)$, contradicting the hypothesis that e' is not in $D(e)$.

Theorem 1.14

Let e be a chord of a spanning tree T in a graph, and f be any edge (other than e) in the fundamental cycle $C(e)$ defined by e. Then: (i) f is an edge of T and e is an edge of the fundamental cutset $D(f)$ defined by f; and (ii) e is not an edge of the fundamental cutset $D(e')$ defined by any edge e' of T which is not in $C(e)$.

Proof

The proof is similar to that of Theorem 1.13.

1.3 MINIMUM WEIGHT SPANNING TREES

Three basic theorems

If w is a mapping from the set of edges of a graph $G = (V, E)$ to the set of real numbers, the graph equipped with the mapping (the weight function) is known as a weighted graph or a network. The number $w(e)$ is called the **weight** of the edge e. An edge e in a set F of edges in the graph is a **minimum weight edge** in the set F if $w(e) \leqslant w(f)$ (or a **maximum weight edge** if $w(e) \geqslant w(f)$) for every f in F. If H is a subgraph of a weighted graph, the weight of H is the sum of the weights of all the edges of H and is denoted by $w(H)$. A spanning tree T in a connected network is a **minimum weight spanning tree** or a **minimal spanning tree** (MST) if there is no other spanning tree T' such that $w(T')$ is less than $w(T)$.

Under what conditions will a spanning tree be a minimum spanning tree? We have two theorems in this context – one involving fundamental cutsets and the other involving fundamental cycles. Loosely speaking, these theorems assert that from the perspective of a minimal spanning tree, every edge in the tree is an edge worthy of inclusion in the tree (cutset optimality condition) whereas every edge not the tree is an edge that can be ignored (cycle optimality condition).

Theorem 1.15
A spanning tree T in a weighted graph is a minimum weight spanning tree if and only if every edge in the tree is a minimum weight edge in the fundamental cutset defined by that edge.

Proof
Let e be an edge of an MST. Suppose $w(e) > w(f)$ for some edge f in the cutset $D(e)$. Consider the spanning tree T' obtained by deleting e from T and adjoining f to it. Then $w(T') < w(T)$, contradicting the fact that T is an MST. Thus the condition is necessary for a spanning tree to be a minimal spanning tree.

On the other hand, suppose T is a spanning tree which satisfies the given condition. Let T' be any minimum weight spanning tree in the graph. If T' and T are not the same, there will be an edge $e = \{i, j\}$ in T which is not an edge in T'.

Let the fundamental cutset $D(e)$ with respect to T be the set of all edges (in the graph) between the vertices in X and the vertices in Y, where $i \in X$ and $j \in Y$. If we adjoin the edge e to the tree T', a unique cycle $C(e)$ is created which contains an edge f (other than e) joining a vertex in X and a vertex in Y.

Now $w(e) \leqslant w(f)$ by our hypothesis. If $w(e) < w(f)$, we can construct a spanning tree T'' by adjoining e to T' and deleting f from it; the weight of T'' is less than the weight of the minimal spanning tree T'. Thus $w(e) = w(f)$. Now if we adjoin the edge f to T and delete e from T we obtain a new spanning tree T_1 such that $w(T) = w(T_1)$. If $T_1 = T'$, then $w(T) = w(T')$, showing that T is an MST. Otherwise we consider an edge in T_1 which is not in T' and repeat the process till we construct a spanning tree T_k such that the weights of T, T_k and T' are all equal.

Corollary 1.16
If e is an edge of any cycle in a connected graph G such that $w(e) \geqslant w(f)$ for every edge f in that cycle, then there is an MST in G which does not contain the edge e. In particular, if $w(e) > w(f)$, no MST in G can have e as an edge.

Proof
Suppose e is an edge in every MST in G and let T be an MST. Since e is an edge in T, according to Theorem 1.15, $w(e) \leqslant w(e')$ for every edge e' in the cutset $D(e)$. Now if we adjoin an edge e' (other than e) from $D(e)$ to T, we obtain a cycle $C(e')$ which contains the edge e.

So according to the hypothesis, $w(e) \geqslant w(e')$. Thus $w(e) = w(e')$. Now if we adjoin e' to T and delete e from T, we obtain a spanning tree T' such that

$w(T') = w(T) + w(e') - w(e) = w(T)$. In other words, T' is an MST which does not contain the edge, contradicting the assumption that e is an edge in every MST. So there is at least one MST which does not contain e.

Suppose the hypothesis is the strict inequality $w(e') < w(e)$. Suppose there exists a minimal spanning tree T which contains e. As above, we can construct a spanning tree T' such that $w(T') = w(T) + w(e') - w(e) < w(T)$, implying that T is not an MST.

Theorem 1.17
A spanning tree T in a graph G is a minimal spanning tree if and only if every chord of the tree is a maximum weight edge in the unique fundamental cycle defined by that edge.

Proof
Let T be a minimal spanning tree in G. Suppose there is a chord e of T such that $w(e) < w(f)$ for some edge f in the cycle $C(e)$. Then the tree T' obtained by adjoining e and deleting f is indeed a spanning tree such that $w(T') = w(T) + w(e) - w(f) < w(T)$, violating the assumption that T is an MST. Thus the condition is necessary.

Let T be a spanning tree in G satisfying the given condition and let f be any edge in this tree T. Now any edge e (other than f) in the fundamental cutset $D(f)$ is a chord of T defining the fundamental cycle $C(e)$, and hence by the hypothesis $w(e) \geqslant w(f)$. This inequality should hold for every e in the cutset $D(f)$. In other words, $w(f) \leqslant w(e)$ for every e in $D(f)$. This is precisely the optimality condition established in Theorem 1.15 and hence T is an MST. Thus the condition is a sufficient condition.

Corollary 1.18
Let e be an edge incident at a vertex x of a connected graph. If $w(e) \leqslant w(f)$ for every other edge f incident at x, then there exists a minimal spanning tree in the graph in which e is an edge. In particular, if $w(e) < w(f)$, every MST in G has e as an edge.

Proof
Suppose $e = \{x, y\}$ is not an edge in any MST. Let T be an MST. Since e is not in T, e is a chord of T. Let $e' = \{x, z\}$ be the other edge in the fundamental cycle $C(e)$ which is incident at the vertex x. In the fundamental cycle $C(e)$, the edge e is a chord and all the other edges are edges in the tree. So by the cycle optimality criterion, $w(e) \geqslant w(f)$ for every f in the cycle $C(e)$. In particular, $w(e) \geqslant w(e')$. But $w(e') \geqslant w(e)$ by our hypothesis. Thus $w(e) = w(e')$. Now the weight of the spanning tree T' obtained by adjoining e to T and deleting e' is $w(T') = w(T) + w(e) - w(e') = w(T)$. Thus there exists a spanning tree T' which contains e, contradicting the assumption.

Suppose $w(e) < w(f)$ and e is not an edge in any MST. Let T be any MST. As before we have the spanning tree T' such that $w(T') = w(T) + w(e) - w(f) < w(T)$, which is a contradiction since T is an MST. Thus e is an edge in every MST.

The following theorem gives us a procedure for constructing a minimal spanning tree in a connected graph G from a subgraph of a known minimal spanning tree in G.

Theorem 1.19

If V' is the set of vertices of a component of any subgraph H of a minimal spanning tree T in $G = (V, E)$ and if e is an edge of minimum weight in the disconnecting set $D = (V', V - V')$, then there exists an MST in G which contains e as an edge and H as a subgraph.

Proof

If e is an edge in T, then the result is obvious. So let us assume that e is a chord in T. In that case $T \cup \{e\}$ has a unique cycle C. The edge e is common to both C and D. So there should be an edge f (other than e) in $(C \cap D)$ since $|C \cap D|$ is even.

Now e is a minimum weight edge in D and f is an edge in D. So $w(e) \leqslant w(f)$.

Every edge in C other than e is an edge in T. In particular, f is an edge in T defining the fundamental cutset $D(f)$. Since $f \in (C \cap D(f))$ and $|(C \cap D(f))|$ is even there should be at least one more edge common to both C and $D(f)$. No edge in $D(f)$ other than f can be an edge in T. The only edge in C which is not in T is the edge e. Thus e is necessarily an edge in $D(f)$.

Since T is an MST, by Theorem 1.15, $w(f) \leqslant w(e)$. Thus $w(e) = w(f)$. If $T^* = T - \{f\} + \{e\}$, then T^* is a spanning tree with weight $w(T^*) = w(T)$. In other words, T^* is an MST in G which contains e as an edge and H as a subgraph.

Two MST algorithms

Using Theorems 1.15, 1.17 and 1.19, we can derive two methods of obtaining an MST in a connected graph. Both procedures are 'greedy' in the sense that at every stage a decision is made to make the best possible choice of an edge for inclusion in the MST without violating any rules. In the course of obtaining a tree the only rule is that at no stage in the decision-making process should the choice of the best possible edge (the 'choicest morsel') for inclusion in an MST produce a cycle.

One greedy procedure (known as **Kruskal's greedy algorithm**) is as follows. The set of edges of the connected graph G with n vertices is listed in nondecreasing weight order. We construct an acylic subgraph T, examining these edges one at a time in the order they are arranged. An edge will be added to T if its inclusion does not yield a cycle. The construction terminates when T has $n-1$ edges. There are four steps in the algorithm:

Step 1. Arrange the edges in nondecreasing order in a list L and set T to be the empty set. (T is the set of edges in an MST.)

Step 2. Add the first edge in L to T.

Step 3. If every edge in L is examined, stop and report that G is not connected. Otherwise take the first unexamined edge in L and include it in T if it does not form a cycle with the edges already in T. If the edge is added to T, go to step 4. Otherwise repeat step 3.

Step 4. Stop if T has $n-1$ edges. Otherwise go to step 3.

The resulting subgraph T must be a spanning tree in G. The fact that T is indeed an MST can be established using Theorem 1.17. Observe that in Kruskal's procedure, an edge e was discarded (in favor of an edge of larger weight) with respect to the tree T because it created a cycle C with some or all of the edges already included in the list L. At the same time, the weight of the discarded edge is greater than or equal to the weight of any other edge in the cycle C because the edges are examined one at a time according to their weights in nondecreasing order. Thus the spanning tree T satisfies the optimality condition as stated in Theorem 1.17.

Another procedure (known as **Prim's greedy algorithm**) to construct an MST is as follows. We start from an arbitrary vertex and add edges one at a time by maintaining a spanning tree T on a subset W of the set of vertices of the graph such that the edge adjoined to T is a minimum weight edge in the cutset $(W, V - W)$. The correctness of this procedure follows directly from the sufficient condition established in Theorem 1.19. There are three steps in this procedure:

Step 1. Select an arbitrary vertex of G and include it in the tree T.

Step 2. Let W be the set of vertices in T. Find an edge of minimum weight in the disconnecting set $(W, V - W)$ and add it to T. If an edge cannot be added, report that G is not connected.

Step 3. Stop if T has $n-1$ edges. Otherwise repeat step 2.

Notice that the procedures described above are applicable even when the graph is not connected. If the input is an arbitrary network G with n vertices, the output will be either an MST with $n-1$ edges or a message that G is not connected.

Example 1.1

Find a minimal spanning tree in the network shown below using (a) Kruskal's algorithm and (b) Prim's algorithm.

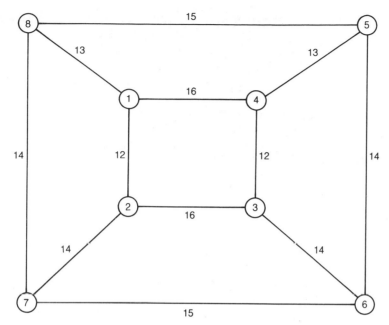

(a) The list of edges with nondecreasing weights is {1, 2}, {3, 4}, {1, 8}, {4, 5}, {7, 8}, {2, 7}, {5, 6}, {3, 6}, {6, 7}, {5, 8}, {2, 3} and {1, 4}, with weights 12, 12, 13, 13, 14, 14, 14, 14, 15, 15, 16 and 16, respectively. We take edges {1, 2}, {3, 4}, {1, 8}, {4, 5} and {7, 8} in that order to form an acyclic subgraph *T*. We discard {2, 7} because the inclusion of this edge in *T* will create a cycle. Then we select {5, 6} and discard {3, 6}. The next edge in the list is {6, 7}, which is included in *T*. At this stage the number of edges in *T* is 7, which is one less than the number of vertices, and we stop. The weight of the MST is $12 + 12 + 13 + 13 + 14 + 14 + 15 = 93$. The graph of this MST is shown below.

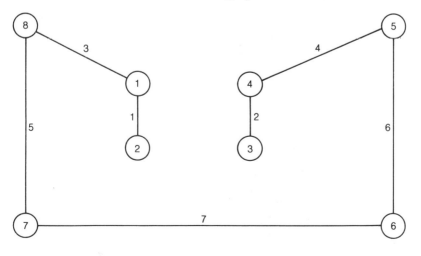

(The positive integer displayed on each edge indicates the order in which it was chosen for inclusion in T, and not the weight of the edge.)

(b) We start from vertex 1, and select the edge $\{1, 2\}$. The set W consists of vertices 1 and 2. An edge of smallest weight in the disconnecting set $(W, V - W)$ is the edge $\{1, 8\}$. At this stage the edges $\{1, 2\}$ and $\{1, 8\}$ are in T and $W = \{1, 2, 8\}$. An edge of smallest weight in $(W, V - W)$ is now either $\{2, 7\}$ or $\{7, 8\}$. Only one of these can be in T since T is acyclic. We include $\{2, 7\}$ and discard $\{7, 8\}$. At this stage $W = \{1, 2, 7, 8\}$, and the edges in T are $\{1, 2\}$, $\{1, 8\}$ and $\{2, 7\}$. We continue like this and include $\{8, 5\}$, $\{5, 4\}$, $\{4, 3\}$ and finally $\{3, 6\}$. At this stage T has 7 edges, and we stop. The weight of the MST is $12 + 13 + 14 + 15 + 13 + 12 + 14 = 93$. The MST obtained by this method is shown below.

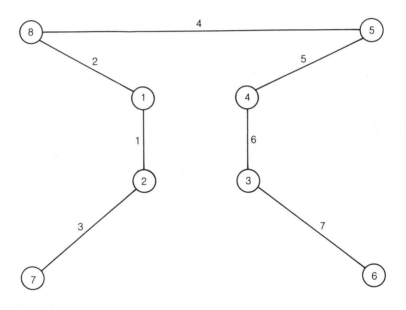

Prim's algorithm (matrix method)

Since a spanning tree in a connected graph passes through every vertex of the graph, we can construct an MST by applying Prim's algorithm starting from any vertex.

In particular if $V = \{1, 2, \ldots, n\}$ is the set of vertices of a connected graph G, we can start from vertex 1. At each stage we have a subtree of the MST. We start from vertex 1 and create a subtree of the MST by adjoining one edge at a time till we use exactly $n - 1$ edges of the graph. We now outline a matrix formulation of this procedure.

Let $\mathbf{D} = [d(i, j)]$ be the $n \times n$ matrix where n is the number of vertices of the graph $G = (V, E)$ and $d(i, j)$ is the weight of the edge $\{i, j\}$ if there is an edge between i and j. Otherwise $d(i, j)$ is plus infinity. Initially delete all elements of column 1 and check row 1 with a check mark $\sqrt{}$. All elements are uncircled initially. Each iteration has two steps as follows.

Step 1. Select a smallest element (ties are broken arbitrarily) from the uncircled entries in the rows which are checked. Stop if no such element exists. The edges which correspond to the circled entries constitute an MST.

Step 2. If $d(i, j)$ is selected in step 1, circle that entry and check row j with a check mark. Delete the remaining elements in column j. Go to step 1.

Example 1.2

Let us consider the network given in Example 1.1. The initial matrix is:

$$
D = \begin{bmatrix}
- & 12 & - & 16 & - & - & - & 13 \\
- & - & 16 & - & - & - & 14 & - \\
- & 16 & - & 12 & - & 14 & - & - \\
- & - & 12 & - & 13 & - & - & - \\
- & - & - & 13 & - & 14 & - & 15 \\
- & - & 14 & - & 14 & - & 15 & - \\
- & 14 & - & - & - & 15 & - & 14 \\
- & - & - & - & 15 & - & 14 & -
\end{bmatrix} \sqrt{}
$$

In this matrix all entries in column 1 are deleted and row 1 is checked. At this stage no entry is circled.

Iteration 1

The smallest uncircled entry in the checked row is $d(1, 2) = 12$. This element is circled, the remaining entries in column 2 are deleted and row 2 is checked, giving the matrix

$$
D = \begin{bmatrix}
- & \circled{12} & - & 16 & - & - & - & 13 \\
- & - & 16 & - & - & - & 14 & - \\
- & - & - & 12 & - & 14 & - & - \\
- & - & 12 & - & 13 & - & - & - \\
- & - & - & 13 & - & 14 & - & 15 \\
- & - & 14 & - & 14 & - & 15 & - \\
- & - & - & - & - & 15 & - & 14 \\
- & - & - & - & 15 & - & 14 & -
\end{bmatrix} \begin{matrix} \sqrt{} \\ \sqrt{} \end{matrix}
$$

Iteration 2

The smallest uncircled entry in the checked rows is $d(1, 8) = 13$. This element is circled, the remaining entries in column 8 are deleted and row 8 is checked, giving the matrix

$$D = \begin{bmatrix} - & \textcircled{12} & - & 16 & - & - & - & \textcircled{13} \\ - & - & 16 & - & - & - & 14 & - \\ - & - & - & 12 & - & 14 & - & - \\ - & - & 12 & - & 13 & - & - & - \\ - & - & - & 13 & - & 14 & - & - \\ - & - & 14 & - & 14 & - & 15 & - \\ - & - & - & - & - & 15 & - & - \\ - & - & - & - & 15 & - & 14 & - \end{bmatrix} \begin{matrix} \checkmark \\ \checkmark \\ \\ \\ \\ \\ \\ \checkmark \end{matrix}$$

Iteration 3

The smallest uncircled entry in the checked rows is $d(2, 7) = 14$. This element is circled, the remaining entries in column 7 are deleted and row 7 is checked, giving the matrix

$$D = \begin{bmatrix} - & \textcircled{12} & - & 16 & - & - & - & \textcircled{13} \\ - & - & 16 & - & - & - & \textcircled{14} & - \\ - & - & - & 12 & - & 14 & - & - \\ - & - & 12 & - & 13 & - & - & - \\ - & - & - & 13 & - & 14 & - & - \\ - & - & 14 & - & 14 & - & - & - \\ - & - & - & - & - & 15 & - & - \\ - & - & - & - & 15 & - & - & - \end{bmatrix} \begin{matrix} \checkmark \\ \checkmark \\ \\ \\ \\ \\ \checkmark \\ \checkmark \end{matrix}$$

Iteration 4

The smallest uncircled entry in the checked rows is $d(7, 6) = 15$. This element is circled, the remaining entries in column 6 are deleted and row 6 is checked, giving the matrix

$$D = \begin{bmatrix} - & \textcircled{12} & - & 16 & - & - & - & \textcircled{13} \\ - & - & 16 & - & - & - & \textcircled{14} & - \\ - & - & - & 12 & - & - & - & - \\ - & - & 12 & - & 13 & - & - & - \\ - & - & - & 13 & - & - & - & - \\ - & - & 14 & - & 14 & - & - & - \\ - & - & - & - & - & \textcircled{15} & - & - \\ - & - & - & - & 15 & - & - & - \end{bmatrix} \begin{matrix} \checkmark \\ \checkmark \\ \\ \\ \\ \\ \checkmark \\ \checkmark \\ \checkmark \end{matrix}$$

Iteration 5

The smallest uncircled entry in the checked rows is $d(6, 5) = 14$. This element is circled, the remaining entries in column 5 are deleted and row 5 is checked, giving the matrix

$$
D = \begin{bmatrix}
- & ⑫ & - & 16 & - & - & - & ⑬ \\
- & - & 16 & - & - & - & ⑭ & - \\
- & - & - & 12 & - & - & - & - \\
- & - & 12 & - & - & - & - & - \\
- & - & - & 13 & - & - & - & - \\
- & - & 14 & - & ⑭ & - & - & - \\
- & - & - & - & - & ⑮ & - & - \\
- & - & - & - & - & - & - & -
\end{bmatrix}
\begin{matrix}
\checkmark \\ \checkmark \\ \\ \\ \checkmark \\ \checkmark \\ \checkmark \\ \checkmark
\end{matrix}
$$

Iteration 6

The smallest uncircled entry in the checked rows is $d(5,4) = 13$. This element is circled, the remaining entries in column 4 are deleted and row 4 is checked, giving the matrix

$$
D = \begin{bmatrix}
- & ⑫ & - & - & - & - & - & ⑬ \\
- & - & 16 & - & - & - & ⑭ & - \\
- & - & - & - & - & - & - & - \\
- & - & 12 & - & - & - & - & - \\
- & - & - & ⑬ & - & - & - & - \\
- & - & 14 & - & ⑭ & - & - & - \\
- & - & - & - & - & ⑮ & - & - \\
- & - & - & - & - & - & - & -
\end{bmatrix}
\begin{matrix}
\checkmark \\ \checkmark \\ \\ \checkmark \\ \checkmark \\ \checkmark \\ \checkmark \\ \checkmark
\end{matrix}
$$

Iteration 7

The smallest uncircled entry in the checked rows is $d(4,3) = 12$. This element is circled, the remaining entries in column 3 are deleted and row 3 is checked, giving the matrix

$$
D = \begin{bmatrix}
- & ⑫ & - & - & - & - & - & ⑬ \\
- & - & - & - & - & - & ⑭ & - \\
- & - & - & - & - & - & - & - \\
- & - & ⑫ & - & - & - & - & - \\
- & - & - & ⑬ & - & - & - & - \\
- & - & - & - & ⑭ & - & - & - \\
- & - & - & - & - & ⑮ & - & - \\
- & - & - & - & - & - & - & -
\end{bmatrix}
\begin{matrix}
\checkmark \\ \checkmark \\ \checkmark \\ \checkmark \\ \checkmark \\ \checkmark \\ \checkmark \\ \checkmark
\end{matrix}
$$

At this stage all the n rows are checked, giving $n-1$ circled entries which correspond to the $n-1$ edges of an MST. These edges are $\{1,2\}$, $\{1,8\}$, $\{2,7\}$, $\{4,3\}$, $\{5,4\}$, $\{6,5\}$ and $\{7,6\}$, giving a total weight of $12 + 13 + 14 + 12 + 13 + 14 + 15 = 93$.

1.4 THE TRAVELING SALESMAN PROBLEM

Optimal Hamiltonian cycles

In general, it is not easy to determine whether an arbitrary graph or digraph is Hamiltonian since there is no known practical characterization of such graphs and digraphs. If a network is Hamiltonian, a problem of interest is to obtain a Hamiltonian cycle of minimum weight, known as an **optimal Hamiltonian cycle**, in the network. Unfortunately there is no efficient procedure to obtain such an optimal Hamiltonian cycle. Finding an optimal Hamiltonian cycle in G by a process of enumeration is a hopelessly inefficient task since the number of Hamiltonian cycles could be very large even when the number of vertices is small. In the parlance of theoretical computer science, we say that the time complexity of (any known procedure to solve every instance of) the problem of finding an optimal Hamiltonian cycle in a network is exponential in the worst case.

The celebrated **traveling salesman problem** (TSP) is the problem of obtaining an optimal Hamiltonian cycle in a connected network if the network is Hamiltonian. A problem of practical importance, however, is the problem of finding a closed path (not necessarily simple) of minimum weight which passes through every vertex at least once in a connected network. Many practical problems in network optimization can be formulated as salesman problems. The real significance of the TSP is not that it has a wealth of applications but that it is a generic problem which captures the essence of several problems in combinatorial optimization.

Even though it is not easy to find an optimal Hamiltonian cycle in a connected graph, we can use the minimal spanning tree algorithm to obtain a lower bound for the weight of such a cycle provided the cycle exists. In this section we assume that the weight $w(e)$ is nonnegative for every edge e in the network under consideration.

Suppose T' is any MST in a network G and T is the spanning tree obtained by deleting an edge from an arbitrary Hamiltonian cycle C. Then $w(T') \leqslant w(T) \leqslant w(C)$. So the weight of an MST in the graph is a lower bound for the weight of an optimal Hamiltonian cycle.

In fact we can obtain a better lower bound using the minimum spanning trees of certain subgraphs of the graph. Suppose C is an optimal Hamiltonian cycle in a graph $G = (V, E)$ with n vertices. If we delete a vertex v and the two edges e and f incident to v in the cycle C, we obtain a spanning tree in the subgraph $G' = (V', E')$ where $V' = V - \{v\}$ and $E' = E - \{e, f\}$.

Suppose T' is an MST in G' with weight $w(T')$. If p and q are two edges incident at the vertex v, of smallest possible weights, then $w(T') + w(p) + w(q)$ is a lower bound for $w(C)$, the weight of the optimal Hamiltonian cycle. Since the graph has n vertices, there will be at most n such (not necessarily distinct) lower bounds. The largest of these lower bounds is a lower bound for the weight of the optimal Hamiltonian cycle if such a cycle exists.

Example 1.3
Find a lower bound for the weight of the optimal Hamiltonian cycle in the network of Example 1.1.

Since the weight of an MST is 93, if the graph has a Hamiltonian cycle its weight cannot be less than 93. If we delete vertex 1 and the three edges incident at this vertex, the weight $w(T')$ of an MST in the subgraph G' is 82. The two edges with least weights incident at vertex 1 have weights 12 and 13. Thus $82 + 12 + 13 = 107$ is a lower bound for the weight of an optimal Hamiltonian cycle in the graph. We obtain the same lower bound if we delete any one of the vertices from the set $W = \{1, 4, 5, 6, 7, 8\}$. But if we delete either vertex 2 or vertex 3, the lower bound is $80 + 12 + 14 = 106$. Thus if a Hamiltonian cycle exists its weight cannot be less than 107.

A quick method to obtain an approximate solution

We now turn our attention to the following question. Suppose it is known that a Hamiltonian cycle exists in a given network G and thus an optimal Hamiltonian cycle C^* with weight $w(C^*)$ exists in G. Is it possible (without too much computational work) to obtain a Hamiltonian cycle C in G such that the gap $w(C) - w(C^*)$ is not too wide? Since an optimal Hamiltonian cycle is a 'lowest weight' Hamiltonian cycle, any Hamiltonian cycle whose weight does not exceed the lowest weight by too much may be called a **low weight Hamiltonian cycle**. For this purpose we restrict our attention to the class of complete graphs with an additional property known as the **triangle property**: if i, j and k are any three vertices in a complete graph and if the weights of the edges $\{i, j\}$, $\{j, k\}$ and $\{k, i\}$ are a, b and c, then $a + b$ cannot be less than c. We then have a theorem which gives an upper bound for this gap.

Theorem 1.20
If a complete network satisfies the triangle property, there exists a Hamiltonian cycle such that its weight is less than twice the weight of the optimal Hamiltonian cycle.

Proof
We prove this by actually constructing a Hamiltonian cycle C in the network with the desired property.

Step 1. Choose any vertex as the initial cycle C_1 with one vertex.
Step 2. Let C_k be a cycle with k vertices. Find the vertex w_k which is not in C_k that is closest to a vertex v_k in C_k.
Step 3. Let C_{k+1} be the new cycle obtained by inserting w_k just prior to v_k in C_k.
Step 4. Repeat steps 2 and 3 till a Hamiltonian cycle C is found.

We now prove $w(C) \leqslant 2w(C^*)$, where C^* is any optimal Hamiltonian cycle in G. Suppose the vertices are $1, 2, 3, \ldots, n$. Assume without loss of generality that the cycle consisting of the edges $\{1, 2\}$, $\{2, 3\}, \ldots, \{n-1, n\}$, $\{n, 1\}$ is an optimal Hamiltonian cycle C^*. Then the path $1 - 2 - 3 - \cdots - n$ is a Hamiltonian path P with weight $w(P)$ which is at most equal to $w(C^*)$.

Our initial cycle C_1 consists of vertex 1 and no edges. With this cycle we associate the set E_1 of all the edges in the Hamiltonian path P.

The weight of the edge $\{i, j\}$ is denoted by $w(i, j)$ or $w(j, i)$.

To construct C_2, with two vertices, we choose the vertex (denoted i) which is closest to vertex 1 in the network. C_2 thus has path 1—i—1, with weight $2w(1, i) \leqslant 2w(1, 2)$. Since we selected the edge $\{1, i\}$ and did not select $\{1, 2\}$, we delete the edge $\{1, 2\}$ from the path P. Thus with cycle C_2 we associate $S_2 = \{\{2, 3\}, \{3, 4\}, \ldots, \{n-1, n\}\}$.

Let j be the vertex not on C_2 closest to a vertex in C_2. If the edge of smallest weight incident at vertex 1 is $e = \{1, p\}$ and the edge of smallest weight incident at i is $f = \{i, q\}$ and if $w(e) < w(f)$, we take $p = j$, C_3 is given by 1—i—j—1. If $w(e) > w(f)$, we take $q = j$ so that C_3 has path 1—j—i—1. (In the case of equality we take either p or q.)

Now

$$w(C_3) - w(C_2) = [w(i, j) + w(1, j) + w(1, i)] - [2w(1, i)]$$
$$= w(i, j) + w(1, j) - w(1, i)$$

But $w(1, j) \leqslant w(i, j) + w(1, i)$ because of the triangle property. Thus

$$w(1, j) - w(1, i) \leqslant w(i, j)$$

Hence

$$w(C_3) - w(C_2) \leqslant 2w(i, j)$$

We choose the edge $\{i, j\}$ for inclusion in the updated cycle C_3. At this stage we locate the first edge in the path P from i (the new vertex in C_2) to j (the new vertex in C_3) and delete this edge from the set S_2 to obtain S_3 which is associated with C_3. This deleted edge is of the form $\{i, i+1\}$ or $\{i, i-1\}$. Since $w(i, j)$ cannot exceed the weight of the deleted edge, $w(C_3) - w(C_2)$ is less than or equal to twice the weight of the edge deleted from P. Proceeding like this, we have $w(C_k) - w(C_{k-1})$ less than or equal to twice the weight of a unique edge from P. Let C_n be denoted by C. Hence

$$w(C) - w(C_{n-1}) \leqslant 2w_{n-1}$$
$$w(C_{n-1}) - w(C_{n-2}) \leqslant 2w_{n-2}$$
$$\ldots$$
$$w(C_3) - w(C_2) \leqslant 2w_2$$
$$w(C_2) \leqslant 2w_1$$

where $w_1, w_2, \ldots, w_{n-1}$ are the weights of the $n-1$ edges of the path P. Addition yields

$$w(C) \leqslant 2(w_1 + w_2 + \cdots + w_{n-1}) = 2w(P) \leqslant 2w(C^*)$$

Thus we obtain a Hamiltonian cycle such that $w(C) \leqslant 2w(C^*)$ and so the gap $w(C) - w(C^*)$ cannot exceed $w(C^*)$. Incidentally, we also obtain $\frac{1}{2}w(C)$ as a lower bound for $w(C^*)$.

Example 1.4

The 5×5 matrix $[w(i, j)]$ given by

$$[w(i, j)] = \begin{bmatrix} 0 & 3 & 3 & 2 & 7 \\ 3 & 0 & 3 & 4 & 5 \\ 3 & 3 & 0 & 1 & 4 \\ 2 & 4 & 1 & 0 & 5 \\ 7 & 5 & 4 & 5 & 0 \end{bmatrix}$$

represents a complete network $G = (V, E)$, where $V = \{1, 2, 3, 4, 5\}$ and $w(i, j)$ is the weight of the edge joining the vertices i and j. It can be verified that G satisfies the triangle property. Use the algorithm described in the proof of Theorem 1.20 to construct a Hamiltonian cycle the weight of which does not exceed twice the weight of any optimal Hamiltonian cycle.

We start from vertex 1. The vertex nearest to 1 is 4. So C_2 is the cycle $1 — 4 — 1$.

The vertex nearest to 4 (not in C_2) is 3, with weight $w(4, 3) = 1$. The vertex nearest to 1 (not in C_2) is 2 or 3 with weight $w(1, 2) = w(1, 3) = 3$. Let us arbitrarily select vertex 3. Thus C_3 is the cycle $1 — 3 — 4 — 1$.

The weights of edges connecting vertices (not in C_3) closest to 1, 3 and 4 are 3, 3 and 4, respectively. The edge of minimum weight 3 is the one joining 3 and 2 or the one joining 1 and 2. We take the edge $\{3, 2\}$. Thus C_4 is the cycle $1 — 2 — 3 — 4 — 1$.

Finally, C_5 (or C) is the cycle $1 — 2 — 5 — 3 — 4 — 1$. Thus C is a Hamiltonian cycle the weight of which cannot exceed twice the weight of any optimal Hamiltonian cycle in G.

Another lower bound for an optimal Hamiltonian cycle

We now describe another algorithm for obtaining a low weight Hamiltonian cycle C in a complete graph $G = (V, E)$ which satisfies the triangle property, using an MST in the graph. Suppose $T = (V, E')$ is an MST in G. Let $H = (V, F)$ be the multigraph obtained by duplicating all the edges of T. Then it can be proved that H has an Eulerian circuit and so it is possible to start from any vertex and return to that vertex by using each edge of H exactly once.

We start from any arbitrary vertex in T and conduct the following DFS of the vertices of the tree which is a systematic method of visiting all the vertices of the graph. We move along the edges of H (using each edge exactly once) from vertex to vertex, assigning labels to each vertex, without revisiting any vertex as long as possible. When a vertex is reached, each of whose adjacent vertices had been already visited, the search revisits the vertex v that was visited immediately before the previous visit to v. This procedure creates a circuit C' in H which uses each edge of H exactly once. Thus $w(C') = w(H) = 2w(T)$. Suppose the vertices in C' appear in the order $v_1, v_2, \ldots, v_n$, which is a permutation of the n vertices of the graph.

Now consider the cycle C'' given by $v_1 - v_2 - v_3 - \cdots - v_n - v_1$. This Hamiltonian cycle may have edges which are not edges in the circuit C'. But by the triangle property, $w(C'') \leqslant w(C')$. Thus $w(C'') \leqslant 2w(T) \leqslant 2w(C)$, where C is any Hamiltonian cycle in the graph. Thus we are able to construct a Hamiltonian cycle C'' in the graph, the weight of which does not exceed twice the weight of any optimal Hamiltonian cycle in the graph.

The algorithm has three steps:

Step 1. Find a minimal spanning tree in the graph.
Step 2. Conduct a depth-first search of the vertices of T.
Step 3. If the order in which the vertices appear in the search is $v_1, v_2, \ldots, v_n$, then the Hamiltonian cycle $v_1 - v_2 - v_3 - \cdots - v_n - v_1$ is a low weight Hamiltonian cycle.

(Since we can start the DFS (in a fixed MST) from any vertex, it may happen that the same tree T may yield more than one low weight Hamiltonian cycle. It is also possible that there will be more than one MST in a given graph.)

Example 1.5
Obtain a low weight Hamiltonian cycle in the complete graph of Example 1.4 using the algorithm described above.

In this graph T is an MST with edges $\{1,2\}, \{1,4\}, \{4,3\}$ and $\{3,5\}$. Suppose we conduct a DFS in this tree starting from vertex 4. The circuit consists of the edges $\{4,1\}, \{1,2\}, \{2,1\}, \{1,4\}, \{4,3\}, \{3,5\}, \{5,3\}$ and $\{3,4\}$. Notice that these are the edges in the multigraph H by duplicating the edges of the minimal spanning tree T. The sequence of vertices in this circuit is 4, 1, 2, 1, 4, 3, 5, 3, 4 which gives the sequence 4, 1, 2, 3, 5 of distinct vertices in the tree. Thus we have a low weight Hamiltonian cycle $4-1-2-3-5-4$ with weight $2+3+3+4+5=17$. The weight of this Hamiltonian cycle also cannot exceed twice the weight of any optimal Hamiltonian cycle in the graph.

1.5 MINIMUM WEIGHT ARBORESCENCES

A digraph is a **directed forest** if its underlying graph is a forest. A **branching** is a directed forest in which the indegree of each vertex is at most 1. A vertex in a branching is a **root** if its indegree is 0. An **arborescence** is a branching which has exactly one root. If a subgraph $T = (V, A')$ of a digraph $G = (V, A)$ is an arborescence with its root at the vertex r, then T is an **arborescence rooted at** r in the digraph G.

A subdigraph H of a digraph G is a **directed spanning tree rooted at** v if H is an arborescence in G with root at vertex v. If a directed spanning tree rooted at v is obtained as a result of a breadth-first search (depth-first search), then it is a BFS (DFS) directed spanning tree rooted at v. If a digraph is not strongly connected, a BFS or DFS starting from an arbitrary vertex v may not result in an arborescence rooted at that vertex.

Obviously, if a digraph is strongly connected it has an arborescence rooted at every vertex. But strong connectivity is not a necessary condition for the existence of an arborescence in a digraph. For example, the digraph $G = (V, A)$, where $V = \{1, 2, 3\}$ and $A = \{(1, 2), (2, 3), (2, 1)\}$, has an arborescence rooted at vertex 1 even though the graph is not strongly connected. We now seek to obtain a sufficient condition to be satisfied by a digraph so that it will have an arborescence.

A digraph is **quasi-strongly connected** if for every pair of vertices u and v in the digraph, there exists a vertex w such that there are directed paths from w to u and from w to v. (The vertex w could be u or v.) Strong connectivity obviously implies quasi-strong connectivity. But the converse is not true, as can be seen from the example given above. If a digraph has an arborescence, then it is obviously quasi-strongly connected. It turns out that quasi-strong connectivity is also a sufficient condition for the existence of a spanning arborescence in a digraph.

Theorem 1.21
A digraph has an arborescence if and only if it is quasi-strongly connected.

Proof
If there is an arborescence in G, then G is quasi-strongly connected. So it is a necessary condition.

On the other hand, suppose $G = (V, A)$ is a quasi-strongly connected digraph, where $V = \{1, 2, 3, \dots, n\}$. Let $H = (V, A')$ be a maximal quasi-strongly connected subgraph of G so that the deletion of one more arc from A' will destroy the quasi-strong connectivity of H.

There exists a vertex x_1 in V such that there are paths (in H) from x_1 to the vertices 1 and 2.

There exists a vertex x_2 such that there are paths (in H) from x_2 to the vertices 3 and x_1.

Finally, there exists a vertex x_{n-1} such that there are paths (in H) from x_{n-1} to vertices n and x_{n-2}. In other words, there exists a vertex v such that there are paths in H from v to every vertex.

Let i be any other vertex in H. Since there is a directed path from v to i, the indegree of i is at least 1. Suppose the indegree of i is more than 1. Then there are at least two distinct vertices j and k such that both (j, i) and (k, i) are arcs in H. So there are at least two paths from v to i. If one of the arcs (j, i) or (k, i) is deleted, then the quasi-strong connectivity of H is not affected. This contradiction shows that the indegree of i is 1.

If there is an arc directed to v, the deletion of this arc will not affect the quasi-strong connectivity of H. Thus the indegree of v is 0. So H is an arborescence in G rooted at v. This completes the proof.

Suppose it is known that a weighted digraph has an arborescence rooted at a vertex r of the digraph. The minimum weight arborescence problem is that of finding an arborescence of minimum weight rooted at r. In the remaining part of this section we consider weighted digraphs. We assume that all arcs in a digraph have different weights.

Suppose $G = (V, A)$ has an arborescence rooted at r. If we adopt the greedy procedure and choose an arc of minimum weight directed to each vertex other than the root, and if the resulting subgraph H is acyclic, then H is indeed an arborescence rooted at r. For example, in the digraph $G = (V, E)$ where $V = \{1, 2, 3, 4\}$ and $A = \{(1, 2), (1, 3), (2, 4), (3, 2), (4, 3)\}$ with weights 3, 5, 4, 6, 7, respectively, this greedy procedure to obtain a minimum weight arborescence rooted at vertex 1 chooses the arcs $(1, 2)$, $(2, 4)$ and $(1, 3)$, giving a minimum weight arborescence in G with a total weight of $3 + 4 + 5 = 12$. But if the arc weights are 6, 7, 4, 3, 5, respectively, the resulting subgraph is no longer acyclic.

Thus the problem is to obtain a procedure to obtain a minimum weight arborescence rooted at a vertex r when the subgraph obtained by the greedy method is not acyclic. The procedure described here is based on the treatment of the same topic by Gondran and Minoux (1984).

Theorem 1.22

Let $T^* = (V, A^*)$ be a minimum weight arborescence rooted at v in the weighted digraph $G = (V, A)$ with distinct arc weights, and let $H = (V, A')$ be the subgraph obtained from G by choosing the arc of minimum weight directed to each vertex other than the root v. Then the set $C - A^*$ has exactly one arc for every cycle C in H.

Proof

Obviously, $C - A^*$ is not empty. Suppose $(C - A^*) = \{e', f'\}$ for some cycle C in H. Let $e' = (p, q)$ and $f' = (r, s)$.

In T^* there is a unique directed path from the root v to q in which the last arc is not (p, q). Let this arc be $e = (p', q)$. Similarly, there is a unique directed path in T^* from the root v to the vertex s the last arc of which is not (r, s) but the arc $f = (r', s)$. The weight of any arc e is $w(e)$. Since the arcs have distinct weights, by our construction of the subgraph H, $w(e') < w(e)$ and $w(f') < w(f)$. Let A'' be the set of arcs obtained from A^* by replacing e by e'. Then $T' = (V, A'')$ is an arborescence rooted at v the weight of which is less than the weight of T^*. But T^* is an arborescence with minimum weight. If more than two arcs of C are not in A^* we arrive at the same contradiction. Thus exactly one arc of C is in A^*.

Construction of a condensed digraph from G

Suppose $G = (V, A)$ has an arborescence rooted at v. Let $H = (V, A')$ be the subgraph obtained by the greedy method. Suppose C is a cycle in H. Let $T^* = (V, A^*)$ be a minimum weight arborescence rooted at r. Let $C - A^*$ be the arc $e' = (k, j)$ as in Theorem 1.22. So there is a unique arc $e = (i, j)$ in A^*, where i is not a vertex in the cycle C.

Let $G_1 = G/C = (V_1, A_1)$ be the graph obtained by shrinking all the vertices of C into a single vertex. Each arc in A_1 corresponds to a unique arc in A. So A_1 can be considered as a subset of A. Then $T_1^* = (V_1, A^* \cap A_1)$ is an arborescence in G_1. We have $w(T_1^*) = w(T^*) + w(C) - w(e')$.

We now define a weight function w_1 on the set A_1. If e is an arc which is not incident to a vertex in C, we define $w_1(e) = w(e)$. Otherwise $w_1(e) = w(e) - w(e')$. Thus w_1 is fixed once C is fixed. Moreover, $w_1(T_1^*) = w(T_1^*) - w(e')$. Thus $w(T^*) - w_1(T_1^*) = w(C)$. Hence T^* is a minimum weight arborescence in G if and only if T_1^* is a minimum weight arborescence in a condensed graph G_1.

Thus we have the following algorithm to obtain a minimum weight arborescence.

(a) Step $i = 0$. $G^0 = G$, $w^0(e) = w(e)$ for each e in the graph G.
(b) At step i, using the weight function w^i, construct the subgraph H^i of G^i by selecting the arc of smallest weight directed to every vertex (other than the root) of G^i.
(c) If there is no directed circuit in H^i, then H^i is an arborescence in G^i from which a minimum weight arborescence of G^0 can be derived. Otherwise go to (d).
(d) If H^i has a directed circuit C, then define $G^{i+1} = G^i/C$ and $w^{i+1}(e) = w^i(e)$ for $e = (i, j)$ where $j \notin C$, and $w^{i+1}(e) = w^i(e) - w^i(e')$ for $e = (i,j)$ where $j \in C$ and $i \notin C$ and $e' = (k, j)$ is an arc in C. (If H^i contains cycles $C_1, C_2, \dots, C_t$, first condense G^i with respect to C_1 to obtain G^i/C_1, then condense G^i/C_1 with respect to C_2, and so on.) Set $i = i + 1$ and return to (b).

If we assume at each step of the algorithm that all arcs of the current digraph have different weights, then the minimum weight arborescence is unique. When there are some arcs with equal weights, the algorithm remains valid since the difference $w(T^*) - w_1(T_1^*)$ is still equal to the weight of the condensed cycle. If C has two arcs $e' = (p, q)$ and $f' = (r, s)$ which are not in A^*, then there are arcs e and f in A^* directed to q and s, respectively. Then $w_1(e) = w(e) - w(e')$ and $w_1(f) = w(f) - w(f')$.

Example 1.6
Obtain a minimum weight arborescence rooted at vertex A in the directed network G^0 shown below:

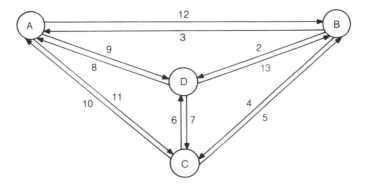

Step 0. The arcs of the digraph H^0 are (B, D), (B, C) and (C, B), with weights 2, 4 and 5, respectively, as shown below:

Since (B, C) and (C, B) form a directed cycle, the vertices B and C are combined to form a new vertex BC. At this stage we construct a condensed graph G^1 with vertices A, BC and D. In this graph the weights of (A, D) and (D, A) are unaffected, being 9 and 8, respectively. The arc in G^0 from A to B with weight 12 becomes an arc from A to BC with weight $12 - 5 = 7$ in the condensed graph. The arc from A to C with weight 11 becomes an arc from A to BC with weight $11 - 4 = 7$.

Since there are two arcs from A to BC we take the one with the smallest weight. As there is a tie it is broken arbitrarily. We find two arcs from BC to A (with weights 10 and 3), two arcs from BC to D (with weights 2 and 6) and two arcs from D to BC (with weights $13 - 5 = 8$ and $7 - 4 = 3$). To avoid multiple arcs we take the arc with the smallest weight in each case. The condensed graph G^1 is constructed as shown below:

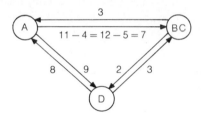

Step 1. We construct H^1 as shown below:

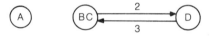

The vertices D and BC constitute a directed cycle. When these two vertices are combined to form a new vertex BCD, we obtain two arcs from A to BCD with weights $7 - 3 = 4$ and $9 - 2 = 7$. If we ignore the arc with the larger weight, the condensed graph G^2 is as given below:

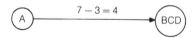

Step 2. At this stage we have a minimum weight arborescence in G^2 which is G^2 itself. We now derive a minimum weight arborescence in G by working backwards from G^2.

Since the weight of the arc (A, BCD) is $7 - 3 = 4$, we locate the arc from A with weight 7 in G^1. This arc is from A to BC. Thus the arborescence is

$A \rightarrow BC \rightarrow D$. The weight of the arc (A, BC) is 7, which is either $11 - 4$ or $12 - 5$. In the former case we take the arcs (A, C) and (C, B). In the latter case we take the arcs (A, B) and (B, C). In G^1, the weight of the arc from BC to D is 2, and this is precisely the weight of the arc from B to D in G. So the arc (B, D) is in the arborescence.

Thus we have two minimum weight arborescences rooted at the vertex A, as shown below. Their arcs are (i) $(A, B), (B, C), (B, D)$; and (ii) $(A, C), (C, B), (B, D)$. he weight of a minimum weight arborescence rooted at vertex A is $12 + 4 + 2 = 11 + 5 + 2 = 18$.

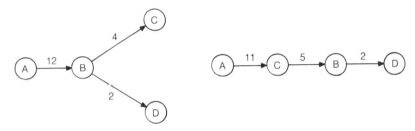

The following observations should be noted. First, if at each stage of the algorithm the current graph G^i has no multiple arcs, then the minimum weight arborescence rooted at vertex v is unique. This is not the case in Example 1.6, where we have chosen arcs of minimum weight whenever multiple arcs appear in the successive condensation processes. Thus a minimum weight arborescence need not be unique even if the weights of the arcs are distinct.

Second, suppose $G = (V, A)$ is a weighted digraph which is quasi-strongly connected. Construct a new vertex r, draw arcs from r to each vertex in V. Let the weight of these new arcs be W, where W is some number larger than the maximum of the set of all weights of the arcs in G. Thus we have a quasi-strongly connected weighted network G' for which G is a subgraph. The problem of finding a minimum weight arborescence in G (without specifying a vertex) is equivalent to the problem of finding a minimum weight arborescence in G' rooted at the vertex r.

1.6 MAXIMUM WEIGHT BRANCHINGS

An acyclic subgraph B of a digraph G is a branching in G if the indegree (in B) of each vertex is at most 1. If w is a real-valued weight function defined on the set of arcs of the digraph, an optimization problem of interest is that of finding a branching B in G such that the sum $w(B)$ of the weights of all the arcs in B is as large as possible. This problem is known as the maximum weight branching problem.

An attempt to solve the maximum weight branching problem by the greedy method need not be successful, as can be seen from the following example. Consider $G = (V, A)$, where $V = \{1, 2, 3, 4\}$ and A is the set $\{(1, 2), (2, 3), (3, 4), (4, 3)\}$ with weights 9, 8, 5 and 10, respectively. The greedy method will give a branching with weight $10+9=19$. But the weight of the maximum weight branching is $9+8+5=22$.

In this section we obtain a procedure for solving the maximum weight branching problem. The algorithm is known as **Edmonds' branching algorithm**. The discussion here is based on a combinatorial proof of this algorithm by Karp (1971).

If $e=(i,j)$ is an arc in the digraph $G=(V,A)$, then the vertex i which is the **source** of e is denoted by $s(e)$. Likewise, the vertex j which is the **terminal** of e is denoted by $t(e)$. Thus both s and t are mappings from A to V. An arc e from vertex p to vertex q in a digraph $G=(V,A)$ with weight function w is a **critical arc** if $w(e)>0$ and $w(e')\leqslant w(e)$ for every arc e' where $t(e')=t(e)$. A spanning subgraph $H=(V,A')$ of G is a **critical graph** if each arc in A' is critical and the indegree of each vertex is at most 1. No two cycles in a critical graph can have an edge or a vertex in common.

If a critical graph $H=(V',A')$ in a digraph G is acyclic, then obviously H is a maximum weight branching. For example, consider the digraph with $V=\{1,2,3,4\}$ and arcs (1, 2), (1, 4), (2, 3), (3, 4) and (4, 2), with weights 6, 5, 3, 2 and 4, respectively. The critical graph is $H=(V,A')$ where the arcs in A' are (1, 2), (1, 4) and (2, 3). H is acyclic and is a maximum weight branching. On the other hand, if the arc weights are respectively 3, 2, 9, 6 and 8, then the critical graph H is (V,A'), where the arcs in A' are (2, 3), (3, 4) and (4, 2). In this case H is not acyclic.

Thus the crux of the problem is to obtain an algorithm to solve the maximum weight branching problem when the critical subgraph is not acyclic. The procedure is as follows. Each cycle in the critical graph is replaced by a single vertex and the weight function is appropriately redefined. We continue this process till we get an acyclic critical graph in which a maximum weight branching is easily discernible. Once we obtain a maximum weight branching in this acyclic graph, we unravel the cycles and revert to the original problem and obtain a maximum weight branching in the given digraph.

Construction of a condensed digraph

Let $G=(V,A)$ be a digraph with weight function w, and H be a critical graph in G. Suppose the cycles in H are C_i $(i=1,2,\ldots,k)$. Let $W=\{v\in V:v$ is not a vertex in any of these k cycles$\}$. Replace each cycle C_i by a vertex X_i. Let $V_1=\{X_1,X_2,\ldots,X_k\}\cup W$. If e is an arc of the digraph G which is not an arc of C_i, and if $t(e)$ is a vertex of C_i, then we define $w_1(e)=w(e)-w(f)+w(e_i)$, where f is the unique arc in C_i with $t(f)=t(e)$ and e_i is an arc of minimum weight in the cycle C_i. If $t(e)$ is not a vertex in any of these k cycles, then $w_1(e)=w(e)$. The multidigraph G_1 thus obtained from H with V_1 as the set of vertices is called the **condensed graph** of the weighted graph. A critical graph in G_1 is denoted by H_1. Thus from the pair (G,H) we move to the pair (G_1,H_1). We continue this process till we reach the pair (G_m,H_m), where H_m is acyclic.

Example 1.7

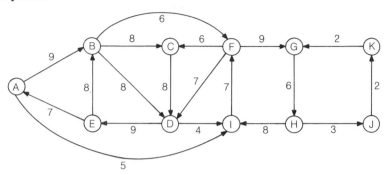

In the digraph G given above, a critical graph in G is the subgraph H as shown below, with two disjoint cycles. The weights of the smallest arcs in these two cycles are 7 and 6, respectively.

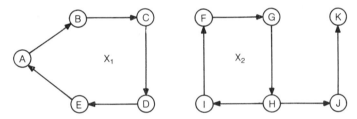

If we replace these two cycles by two vertices X_1 and X_2 and readjust the weight function, we get the condensed graph G_1 shown below. There are three arcs from X_1 to X_2 corresponding to the arcs (A, I), (B, F) and (D, I) with revised weights $5-8+6=3$, $6-7+6=5$ and $4-8+6=2$, respectively. There are two arcs from X_2 to X_1 corresponding to (F, C) and (F, D) with revised weights $6-8+7=5$ and $7-8+7=6$.

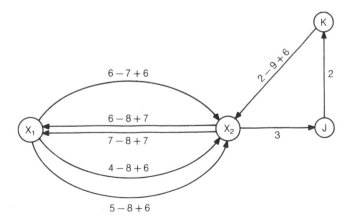

The arc (H, J) with weight 3 is replaced by the arc (X_2, J) with the same weight. The arc (K, G) is replaced by (K, X_2) with no change in weight. The arc (J, K) remains unchanged. A critical graph in G_1 is H_1 as shown below.

The cycle in H_1 is condensed into a single vertex X_{12} to form the following graph G_2, which is acyclic.

We now turn our attention to the theoretical justification behind this procedure as well as the unraveling process in which the condensed cycles are expanded. Let $B = (V', A')$ be a branching in a digraph $G = (V, A)$. An arc e in A which is not in A' is called a B-**eligible arc** if the arcs in the set

$$A'' = A' \cup \{e\} - \{f \in A': t(e) = t(f)\}$$

constitute a branching. Observe that there is at most one arc in the set $\{f \in A': t(e) = t(f)\}$. Eligibility is on an individual basis. If e is B-eligible and f is B-eligible it is not necessary that both e and f are simultaneously B-eligible.

Theorem 1.23
An arc e in a digraph is B-eligible if and only if there is no directed path in the branching B from $t(e)$ to $s(e)$.

Proof
Let $B = (V', A')$ be a branching in $G = (V, A)$. The set

$$A'' = A' \cup \{e\} - \{f \in A': t(e) = t(f)\}$$

will form a branching if and only if it does not contain a cycle since B is a branching. Since B is acyclic, any cycle in A'' should contain the arc e. In other words, C is a cycle in A'' if and only if $C - \{e\}$ is a directed path in B from $t(i)$ to $s(i)$. Thus e is B-eligible if and only if there is no directed path in B from $t(e)$ to $s(e)$.

Suppose $B = (V', A')$ is a branching in $G = (V, A)$. Let C be the set of arcs in a cycle in G. Since C cannot be a subset of A', the set $C - A'$ is nonempty. If $C - A'$ consists of one arc e, then e is not B-eligible. The converse also is true, and this is the content of the next theorem.

Theorem 1.24
If $B = (V', A')$ is a branching in $G = (V, A)$ and if C is the set of arcs in a cycle such that no arc in $C - A'$ is B-eligible, then $C - A'$ contains exactly one arc.

Proof

Suppose $C - A' = \{e_i : i = 1, 2, \ldots, k\}$. Assume that these arcs appear clockwise in the cycle such that e_{i+1} follows immediately after e_i for $i = 1, 2, \ldots, k-1$. Hence $t(e_{i-1}) = s(e_i)$ or there is a path in $C \cap A'$ from $t(e_{i-1})$ to $s(e_i)$ for $i = 2, 3, \ldots, k$. Also, either $t(e_k) = s(e_1)$ or there is a path in $C \cap A'$ from $t(e_k)$ to $s(e_1)$.

Suppose no arc in $C - A'$ is B-eligible. Then there exists a path in B from $t(e_i)$ to $s(e_i)$ for each i. Now there is a path in $C \cap A'$ from $t(e_{i-1})$ to $s(e_i)$. There is also a path in A' from $t(e_i)$ to $s(e_i)$. So there is a path in A' from $t(e_{i-1})$ to $t(e_i)$ or there is a path in A' from $t(e_i)$ to $t(e_{i-1})$. In the former case, there is a path $t(e_{i-1}) \to t(e_i) \to s(e_i)$ in the branching A'. But the unique path in A' from $t(e_{i-1})$ to $s(e_i)$ is in $C \cap A'$. So the path from $t(e_{i-1})$ to $t(e_i)$ is also in $C \cap A'$. So there should be an arc in $C \cap A'$ directed to $t(e_i)$. But there is already the arc e_i directed to $t(e_i)$ in the cycle. This arc e_i is not B-eligible and therefore it is not in A'. Thus we have two arcs in the cycle – one in A' and one in its complement – such that both are directed to the vertex $t(e_i)$. So there is no path in A' from $t(e_{i-1})$ to $t(e_i)$ for each i. Hence there is a path in A' from $t(e_i)$ to $t(e_{i-1})$ for $i = 1, 2, \ldots, k$ and there is a path in A' from $t(e_1)$ to $t(e_k)$. This situation creates the following cycle C' in A':

$$t(e_k) \to t(e_{k-1}) \to \cdots \to t(e_1) \to t(e_k)$$

leading to a contradiction since A' is acyclic.

In the figure below, the arcs in the branching B are thick lines and the arcs not in B are dotted lines. The arc $(8, 1)$ is eligible and there is no path from 1 to 8 in B. By adjoining this arc to B, the cardinality of the set A' of arcs in the branching is increased by one. Likewise, the arc $(2, 9)$ is eligible and there is no path from 9 to 2. The arc $(1, 6)$ is B-eligible and if it enters B, the arc $(5, 6)$ has to leave B. The arc $(4, 1)$ is not eligible and there is a path from 1 to 4 in B. The cycle $1 \to 2 \to 9 \to 8 \to 1$ has $(1, 2)$ in the branching and the three remaining arcs are B-eligible. In the cycle $1 \to 2 \to 4 \to 1$, the only arc not in the branching is $(4, 1)$ which is not B-eligible.

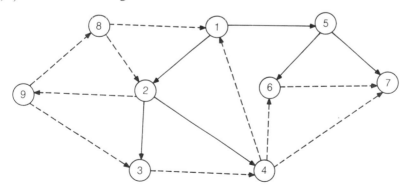

Theorem 1.25

If H is a critical subgraph of a digraph $G = (V, A)$, there exists a maximum weight branching $B = (V, A')$ in G such that, for every cycle C in H, the set $C - A'$ has exactly one arc.

Proof
Let $H = (V, A_1)$ be a critical graph. From the collection of all maximum weight branchings, choose that branching $B = (V, A')$ which has the maximum number of arcs in common with the critical graph H. Let $e = (i, j)$ be any arc in A_1 but not in A'. Suppose e is B-eligible. Then the arcs in

$$A'' = A' \cup \{e\} - \{f : f \in A' \text{ and } t(e) = t(f)\}$$

form a maximum weight branching which has more arcs from the critical graph than A' has. So no arc in H is B-eligible. In particular, no arc in a cycle C in H is B-eligible. Hence by Theorem 1.24, the cardinality of the set $C - A'$ is 1.

Theorem 1.26
Suppose the cycles in a critical subgraph H of a digraph G are $C_1, C_2, \ldots, C_k$. Then there exists a maximum weight branching $B = (V, A')$ in G such that (i) $C_i - A'$ has exactly one arc for each i, and (ii) if arc e is not directed to a vertex in the cycle C_i for every arc e in the set $A' - C_i$, then $C_i - A'$ is an arc of minimum weight in C_i for every i.

Proof
Observe that these cycles are disjoint in the sense that no two cycles have a vertex or an arc in common.

Let e_i be an arc of minimum weight in the cycle C_i. Let $S = \{e_1, e_2, \ldots, e_k\}$. By Theorem 1.25, there exists (not uniquely) a maximum weight branching (which depends on H but not on the individual cycles) such that every arc except one of each C_i is an arc of this branching. Choose a maximum spanning branch $B = (V, A')$ of this kind which contains the minimum number of arcs from the set S. This branching B satisfies (i).

Suppose this branching B does not satisfy property (ii). So there exists a cycle C_j ($1 \leqslant j \leqslant k$) where this property does not hold. No arc of $A' - C_j$ is directed to a vertex in V_j and e_j is not the arc $C_j - B$. So the arc e_j is an arc in the branching B. Let e be the arc $C_j - B$. So $w(e) \geqslant w(e_j)$. Then $A' - (e_j) \cup \{e\}$ is a maximum weight branching which satisfies property (i) and has fewer edges than A' from the set S. This is a contradiction. So B satisfies property (ii). Thus B is the desired maximum weight branching in the digraph.

Theorem 1.27
There is a one-to-one correspondence between the set of maximum weight branchings in a weighted digraph G satisfying properties (i) and (ii) of Theorem 1.26 and the set of maximum weight branchings in a condensed digraph in G.

Proof
Let $G = (V, A)$ be a digraph with a weight function w. Let H be a critical graph in G with cycles C_i $(i = 1, 2, \ldots, k)$ and let $G_1 = (V_1, A_1)$ be the corresponding condensed graph.

We can define a weight function w_1 on this condensed graph by the way the arcs in A_1 are classified. The set A_1 consists of two categories of arc: an arc e is in the first category when $t(e)$ is not a vertex in any of the cycles of H, in which case $w_1(e) = w(e)$. Otherwise the arc e is in the second category in which case $w_1(e) = w(e) - w(f) + w(e_i)$, where f is the unique arc in the cycle C_i which is directed to $t(e)$ and e_i is an arc of minimum weight in C_i.

Let $B = (V, A')$ be any branching in G satisfying properties (i) and (ii) of Theorem 1.26 using these cycles C_i.

An arc e in A' such that both $s(e)$ and $t(e)$ are not in the same cycle in H defines a unique arc in A_1; let D_1 be the set of arcs thus defined. Then $B_1 = (V_1, D_1)$ is a branching in G_1. (We may say that D_1 is the 'intersection' of A' and A_1.) Thus once a critical graph is fixed in G, a branching in G defines a unique branching in the condensed graph G_1 defined by the critical graph.

Now consider the condensed graph $G_1 = (V_1, A_1)$ defined by a critical graph H in a digraph G. Let the cycles in H be C_i $(i = 1, 2, \ldots, k)$. Let $B_1 = (V_1, A_1)$ be a branching in G_1.

If the indegree in B_1 of the condensed vertex corresponding to the cycle C_i is 0, let $C_i' = C_i - \{e_i\}$.

If this indegree is 1, there is a unique arc f (belonging to G) in C_i which is responsible for this. In this case let $C_i' = C_i - \{f\}$. Thus from each cycle in H, we take all the arcs except one. Let X be the set of all arcs thus obtained from the k cycles.

Now consider arcs in B_1 which are directed to vertices which are not condensed vertices. Each such arc corresponds to a unique arc in G. Let Y be the set of arcs in G thus obtained. Then the union of X and Y constitutes a branching B in G.

Now we turn our attention to optimality. Let B and B_1 be as described above. Let P be the sum of the weights of the k cycles in H. Let Q be the sum of the weights of the minimum weight arcs in these k cycles, taking exactly one arc from each cycle. It is a simple exercise to verify that $w(B) - w_1(B_1) = P - Q$.

Thus there is a one-to-one correspondence between the set of maximum weight branchings in a digraph G and the set of maximum weight branchings in a condensed digraph obtained by condensing each cycle in a critical graph H in G into a vertex.

Observe that Theorem 1.27 gives us a procedure for obtaining a maximum weight branching in G. First construct a critical subgraph H. If H is acyclic, then H is a maximum weight branching and we are done. Otherwise, we shrink each cycle into a vertex and readjust the weight function to obtain a condensed graph G_1. We continue this process till we reach an acyclic critical graph H_m of the condensed graph G_m. Thus we move from (G, H) to (G_m, H_m). Now the unraveling starts. H_m gives a maximum weight branching in G_m. We expand the condensed vertices in H_{m-1} and obtain a maximum weight branching in G_{m-1}. We continue this process of expanding condensed cycles till we get a maximum weight branching in the digraph G.

We can summarize the maximum weight branching algorithm as follows:

Step 1 (the condensation process). The input is $G = G_0$. Construct G_i from G_{i-1} by condensing cycles of a critical subgraph H_{i-1} of G_{i-1} and by modifying the weights of the arcs. G_k is the first digraph in the sequence with a critical acyclic subgraph H_k.

Step 2 (the unraveling process). The graph H_k is a maximum weight branching in G_k. Let $B_k = H_k$. Construct B_{i-1} from B_i by expanding the condensed cycles. B_i is a maximum weight branching in G_i for $i = k$, $k - 1, \ldots, 0$. $B = B_0$ is the output.

Example 1.8

Obtain a maximum weight branching in the digraph G of Example 1.7.

In the condensation process, we moved from (G, H) to (G_1, H_1) and then to (G_2, H_2), where $H_2 = G_2$ as shown in Example 1.6. When we open the condensed cycle X_{12} into two vertices X_1 and X_2 we have the maximum weight branching B_1 in the graph G_1 as shown below.

At this stage we open the vertices X_1 and X_2 to obtain the maximum weight branching B in G as follows:

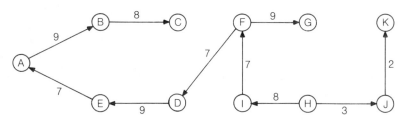

If A' is the set of arcs in the maximum weight branching the set $A' - C_2$ consists of the arcs (A, B), (B, C), (E, A), (D, E), (F, D), (H, J) and (J, K). No arc from this is directed to a vertex in C_2. Consequently the set $C_2 - A'$ should be a singleton set consisting of an arc with the smallest weight in the cycle C_2, as is indeed the case here. On the other hand, when we consider the set $A' - C_1$ there is an arc, namely the arc (F, D), directed to a vertex in the cycle C_1. So the unique arc in $C_1 - A'$ need not be an arc of minimum weight in C_1. Moreover, the sum X of the weights in the two cycles in H is $41 + 30 = 71$ and the sum Y of the two minimum weight arcs is $7 + 6 = 13$. Thus $X - Y = 58$. The weight of the optimal branching in G is 69 and the weight of the optimal branching in the condensed graph H_1 shown in Example 1.7 is $6 + 3 + 2 = 11$, and the difference between the weights of these two branchings is also 58.

In the general case, the graph at the beginning of iteration $i + 1$ is G_i, with the updated weight function w_i. (Initially $G_i = G_0 = G$ and $w_i = w_0 = w$.) Suppose the cycles in G_i are $C_{i,1}, C_{i,2}, \ldots, C_{i,k}$. Let $e_{i,j}$ be an arc of minimum weight in $C_{i,j}$ for each j. If $p_i = \Sigma [w_i(C_{i,j}) - w_i(e_{i,j})]$, where j varies from 1 to k, and if q_{i+1} is the weight of a maximum weight branching in G_{i+1}, then $p_i + q_{i+1}$ is the weight of a maximum weight branching in G_i.

1.7 EXERCISES

1. Show that a graph with exactly one spanning tree is tree.
2. Prove that if G is any connected graph with n vertices with the property that every subgraph of G with $n-1$ edges is a spanning tree, then G is either a tree or a cycle.
3. Show that every edge of a connected graph G is an edge of some spanning tree of G.
4. Show that a subgraph H of a connected graph is a spanning tree if and only if H is a maximal acyclic subgraph of G.
5. Show that every connected graph contains a cutset.
6. Find a minimum weight spanning tree in the network below using:
    (i) Kruskal's algorithm;
    (ii) Prim's algorithm;
    (iii) Prim's algorithm (matrix method).

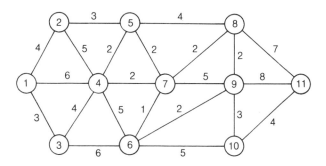

7. Show that if the weights of the edges of a connected network are all unequal, the network has a unique minimum weight spanning tree.
8. Obtain (i) a minimum weight spanning tree, (ii) a maximum weight spanning tree, (iii) a lower bound for an optimal Hamiltonian cycle in the network below.

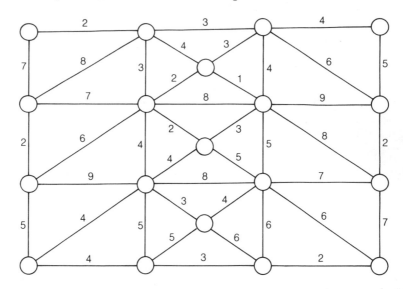

9. If $e = \{i, j\}$ is an edge of the weighted graph $G = (V, E)$, where $V = \{1, 2, \ldots, n\}$ with weight function w, the $n \times n$ matrix $\mathbf{A} = [a_{ij}] = [w(e)]$ is called the weight matrix of G. Suppose the weight matrix of a graph is as follows:

$$\mathbf{A} = \begin{bmatrix} - & 7 & 3 & 2 & 3 & 3 \\ 7 & - & 4 & 5 & 5 & 4 \\ 3 & 4 & - & 5 & 5 & 4 \\ 2 & 5 & 5 & - & 4 & 1 \\ 3 & 5 & 5 & 4 & - & 3 \\ 3 & 4 & 4 & 1 & 3 & - \end{bmatrix}$$

(i) Show that G has a spanning tree.
(ii) Show that G is Hamiltonian.
(iii) Obtain a minimum weight spanning tree in G.
(iv) Obtain a lower bound for the weight of the minimum weight Hamiltonian cycle in G.
(v) Show that G has the triangle property.
(vi) Obtain an approximate optimal Hamiltonian cycle using the procedure in Theorem 1.20.

10. Construct a minimum weight spanning tree in the network of Exercise 6 above that includes the edges $\{1, 2\}$, $\{1, 3\}$ and $\{1, 4\}$.

11. Obtain a minimum weighted arborescence rooted at vertex 1 in the quasi-strongly connected weighted digraph below.

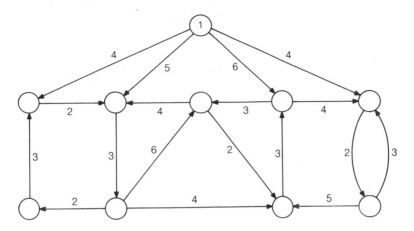

12. Obtain a maximum weighted branching in the weighted digraph in Exercise 11 above.

13. Obtain a maximum weighted branching in the weighted digraph below.

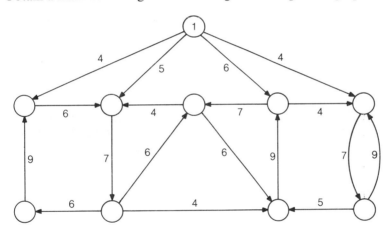

14. Obtain a minimum weighted arborescence rooted at vertex 1 in the digraph in Exercise 13 above.

2

Transshipment problems

2.1 THE NETWORK SIMPLEX METHOD

Statement of the problem

Let $G = (V, E)$ be a digraph with n vertices and m arcs. With each arc (i, j) we associate a real number c_{ij} which is the cost of sending one unit of a prescribed commodity from i to j along the arc (i, j). The row vector $\mathbf{c} = [c_{ij}]$ with m components is called the **cost vector**. Some of the vertices are called **sources**, and these represent where the commodity is produced or manufactured. Those vertices where the commodity is demanded for consumption are called **sinks**. An **intermediate vertex** is one which is neither a source nor a sink.

If i is a source producing b_i units of the commodity the **supply** at i is denoted by $-b_i$. If j is a sink where b_j units are consumed the **demand** at j is denoted by b_j. If k is an intermediate vertex we define $b_k = 0$. The column vector $\mathbf{b} = [b_i]$ with n components in which the ith component is negative, zero or positive depending upon whether i is a source, an intermediate vertex or a sink is called the **supply–demand vector**. In order to simplify our discussions we make an (unrealistic) assumption that the total supply and the total demand of the commodity are the same in the network. This requirement, known as the **equilibrium condition**, will later be relaxed.

With each arc (i, j) we associate a variable nonnegative number x_{ij}, called the **flow** from i to j along that arc, which is the amount of the commodity that is currently being sent along the arc (i, j). The column vector $\mathbf{x} = [x_{ij}]$ is the **flow vector** and the real number z, where $z = \mathbf{cx}$ is called the **cost of the flow**. The total amount of the commodity that is sent to a vertex i is called the **inflow** at vertex i and the total amount that is sent from it is called the **outflow** at i.

We obtain the **netflow** at i by subtracting the outflow at i from the inflow at i. A flow vector $\mathbf{x}$ is said to be **feasible** if the netflow at i is equal to $-b_i$ if i is a source, b_i if i is a sink, or zero otherwise. A network with a supply–demand vector is a **feasible network** if there is at least one feasible flow in that network.

In this chapter and in the following one, we assume that there is no upper bound restriction on the amount of commodity that can be sent from i to

j along (i, j), making the network an uncapacitated network. The **single commodity uncapacitated transshipment problem** is the problem of finding a feasible flow in an uncapacitated feasible network such that its cost is as low as possible.

In our discussion we assume that all components of the cost vector **c** are nonnegative. We also assume that the network is weakly connected.

If **x** is a feasible flow and if r_i is the ith row of the incidence matrix **A** of the network then $r_i\mathbf{x}$ is negative, zero or positive depending upon whether i is a source, an intermediate vertex or a sink. More precisely, **x** is feasible if and only if $\mathbf{Ax} = \mathbf{b}$ and **x** is nonnegative (i.e. each component of **x** is nonnegative).

So our problem can be stated as follows. Given a weakly connected digraph G with n vertices and m arcs, a cost vector **c** with m nonnegative components and a supply–demand vector **b** with n components satisfying the equilibrium condition, determine whether there exists a feasible vector **x** (with m non-negative components such that $\mathbf{Ax} = \mathbf{b}$ (where **A** is the incidence matrix of G), and, if the answer is in the affirmative, obtain a feasible vector **x** such that the total cost **cx** is as small as possible. A feasible vector which minimizes the **objective function** cx is called an **optimal solution** of the problem. Notice that because of the equilibrium condition, one of the n equations (constraints) in the linear system $\mathbf{Ax} = \mathbf{b}$ can be considered as redundant.

Thus we have a **constrained linear optimization problem** (involving a linear objective function of m nonnegative variables and a set of $n - 1$ linear equality constraints) which is a special case of a standard linear programming (LP) problem. A powerful and popular algorithm to solve the LP problem is the simplex method of Dantzig (1963). In this chapter we develop a network-based specialization of the simplex method, known as the network simplex method, to solve the transshipment problem.

Feasible tree solution

If a network is weakly connected, its underlying graph has a spanning tree which defines (when each edge is replaced by the original arc) a spanning tree in the network. A feasible flow vector **x** in the network is called a **feasible tree solution** (FTS) if there is a spanning tree T in the network such that $x_{ij} = 0$ whenever (i, j) is not an arc of T. Thus, corresponding to every FTS in a network with n vertices, there exists an **associated tree** with $n - 1$ arcs. In the context of linear programming, an FTS is what is known as a **basic feasible solution**.

If a problem has an FTS, then we shall show that it has an optimal FTS. This fact allows us to start (initialization) with an arbitrary FTS (basic feasible solution) and see whether it is optimal (optimality test). If it is not optimal, we proceed (pivot operation) to another FTS. We stop as soon as we obtain an optimal FTS (termination). Our discussion here is based on the elaborate treatment of this topic in Chvátal (1983).

If the network has n vertices and if the number of nonzero components in an FTS is $n-1$, then the FTS is **nondegenerate**, in which case the associated tree is unique. A feasible tree solution is **degenerate** if the number of nonzero components is less than $n-1$, in which case the associated tree is not unique. For example, in the network where $V = \{1, 2, 3, 4\}$, $\mathbf{b} = [-3\ 3\ 0\ 0]$ and $E = \{(1, 2), (2, 3), (3, 4), (4, 1)\}$, with flow vector $\mathbf{x} = [x_{12}\ x_{23}\ x_{34}\ x_{41}]$, the solution $[3\ 0\ 0\ 0]$ is an FTS for which the set of arcs of an associated spanning tree could be $\{(1, 2), (2, 3)\}$ or $\{(1, 2)\ (4, 1)\}$. So there is at least one associated tree of an arbitrary feasible tree solution.

On the other hand, if T is an arbitrary spanning tree in a network, there is at most one feasible tree solution for which T can be an associated tree, as shown in the following theorem.

Theorem 2.1
If $T(\mathbf{x})$ and $T(\mathbf{y})$ are spanning trees associated with feasible tree solutions $\mathbf{x}$ and $\mathbf{y}$, respectively, and if $T(\mathbf{x}) = T(\mathbf{y})$, then $\mathbf{x} = \mathbf{y}$.

Proof
Suppose the vertices of T are $\{1, 2, \ldots, n\}$. Then we can relabel these vertices as $\{v_1, v_2, \ldots, v_n\}$ and relabel the arcs as $\{a_2, a_3, \ldots, a_n\}$ according to the following procedure.

Relabel any one of the n vertices as v_1. At each stage i ($i > 1$), there is exactly one arc such that one of its endpoints is v_i and the other endpoint is one of the vertices from $V_i = \{v_1, v_2, \ldots, v_{i-1}\}$. For the arc a_i, one of the endpoints is v_i and the other endpoint is a vertex from the set V_i of vertices already relabeled.

Suppose $\mathbf{B}$ is the $(n-1) \times (n-1)$ matrix obtained by deleting the first row of the incidence matrix of the tree T with vertices $\{v_1, v_2, \ldots, v_n\}$ and arcs $\{a_2, a_3, \ldots, a_n\}$. By the way the vertices and arcs are relabeled, this matrix $\mathbf{B}$ is upper triangular, in which each diagonal entry is either -1 or 1. So $\mathbf{B}$ is nonsingular. This completes the proof.

The dual solution defined by a spanning tree

Corresponding to a given spanning tree T we can define a row vector $[y_1\ y_2 \cdots y_{n-1}\ 0]$ with n components by means of the following **defining equations**:

$$-y_i + y_j = c_{ij} \qquad \text{for each arc } (i, j) \text{ of the tree}$$

We thus have a linear system of $n-1$ equations with $n-1$ variables. The **dual solution** $\mathbf{y}$ is the $1 \times n$ vector whose components are the n components (in the same order) of the row vector defined above. The system of defining equations can then be written as $\mathbf{y}'\mathbf{B} = \mathbf{c}'$, where $\mathbf{c}'$ is a subvector of the cost vector $\mathbf{c}$, $\mathbf{y}'$ is the subvector of $\mathbf{y}$ consisting of the first $n-1$ components of $\mathbf{y}$, and $\mathbf{B}$ is the square matrix obtained by deleting a row (not necessarily the first row) from the incidence matrix of the tree. Now the rows and columns of the incidence matrix can be permuted (by the relabeling process as in Theorem 2.1) such that the matrix **B** is upper triangular with nonzero diagonal entries. So once the tree is fixed, the dual solution of an FTS is unique. Hence the dual solution of a nondegenerate FTS is unique. However, if T and T' are two distinct spanning trees associated with a degenerate FTS giving rise to two dual solutions **y** and **y'**, the vectors **y** and **y'** need not be equal.

Suppose **x** is an FTS defining a unique tree $T(\mathbf{x})$ and a unique dual solution **y**. An arc (i, j) which is not in $T(\mathbf{x})$ is said to be **profitable with respect to the FTS** if $y_j - y_i > c_{ij}$.

Theorem 2.2
A feasible tree solution with no profitable arcs is an optimal solution.

Proof
Let **u** be an FTS associated with the tree $T(\mathbf{u})$ and **y** be the dual solution. Define $\mathbf{c}' = \mathbf{c} - \mathbf{y}\mathbf{A}$, where **c** is the cost vector and **A** is the incidence matrix of the network.

Now if (i, j) is an arc of the tree $T(\mathbf{u})$, then $c'_{ij} = 0$ since

$$c'_{ij} = c_{ij} - (-y_i + y_j) = c_{ij} + (y_i - y_j)$$

If (i, j) is not an arc of the tree, $x_{ij} = 0$. Thus $\mathbf{c}'\mathbf{u} = 0$.

Suppose **v** is any feasible solution, not necessarily a feasible tree solution. So $\mathbf{A}\mathbf{v} = \mathbf{b}$ and **v** is nonnegative. Thus

$$\mathbf{c}\mathbf{v} = (\mathbf{c}' + \mathbf{y}\mathbf{A})\mathbf{v} = \mathbf{c}'\mathbf{v} + \mathbf{y}\mathbf{A}\mathbf{v} = \mathbf{c}'\mathbf{v} + \mathbf{y}\mathbf{b}$$

for any feasible solution **v**. In particular,

$$\mathbf{c}\mathbf{u} = \mathbf{c}'\mathbf{u} + \mathbf{y}\mathbf{b}$$

since **u** is a feasible solution.

But $\mathbf{c}'\mathbf{u} = 0$. So $\mathbf{c}\mathbf{u} = \mathbf{y}\mathbf{b}$ and hence

$$\mathbf{c}\mathbf{v} = \mathbf{c}'\mathbf{v} + \mathbf{c}\mathbf{u}$$

Thus any feasible solution **v** and any feasible tree solution **u** are connected by the relation $\mathbf{c}\mathbf{v} = \mathbf{c}'\mathbf{v} + \mathbf{c}\mathbf{u}$, where $\mathbf{c}' = \mathbf{c} - \mathbf{y}\mathbf{A}$, **y** being the dual solution corresponding to **u**. Hence

$$\mathbf{c}\mathbf{v} - \mathbf{c}\mathbf{u} = \mathbf{c}'\mathbf{v}$$

Suppose the tree has no profitable arc. In that case $\mathbf{c}'\mathbf{v}$ is nonnegative since c'_{ij} is nonnegative (whenever the arc (i, j) is not in the tree) or 0 (whenever the arc is in the tree). So if the tree associated with the feasible tree solution **u** has no profitable arcs then **cu** cannot exceed **cv**, where **v** is any feasible solution. In other words, **u** is an optimal FTS.

It follows from Theorem 2.2 that if **u** is not an optimal FTS then there is at least one arc which is profitable with respect to **u**. In the second stage of the algorithm we select an arbitrary profitable arc e (called the **entering arc**) and adjoin this arc e to $T(e)$ to obtain a unique cycle $C(e)$. An arc in this cycle which has the same direction as e is a **forward** arc and an arc in the opposite direction is **backward**. We send t units of flow (the commodity) along e and at the same time increase the current flow in each forward arc by t units and decrease the current flow in each backward arc by t units, where of course t is nonnegative. We choose t such that the flow in at least one backward arc f becomes 0 and the flow in all other arcs is nonnegative. We do not rule out the possibility that t could be 0. Since the cost vector is assumed to be nonnegative, at least one of the arcs in the cycle is backward. An arc f in $C(e)$ for which the revised flow is 0 is called the **leaving arc**. We now have an updated flow vector **v** associated with the tree $T(\mathbf{v})$ obtained by deleting f from $T(\mathbf{u})$ and adjoining e to it at the same time.

Theorem 2.3
If **u** is an FTS with a profitable arc and if **v** is the updated flow using this profitable arc, then **v** is an FTS and $\mathbf{cv} \leqslant \mathbf{cu}$.

Proof
When we change the feasible flow vector **u** into the flow vector **v** the netflow at each vertex remains unaffected and therefore **v** is feasible. Furthermore, the components of **v** with respect to the arcs not in $T(\mathbf{v})$ are 0. In other words **v** is the FTS associated with $T(\mathbf{v})$.

Let $\mathbf{c'} = \mathbf{c} - \mathbf{yA}$, where **y** is the dual solution of **u**. So $\mathbf{c} = \mathbf{c'} + \mathbf{yA}$, which implies $\mathbf{cv} = \mathbf{c'v} + \mathbf{yb}$ and $\mathbf{cu} = \mathbf{c'u} + \mathbf{yb}$. But $\mathbf{c'u} = 0$ as proved in Theorem 2.2. So it is enough to prove that $\mathbf{c'v} \leqslant 0$.

If (i, j) is any arc of the tree $T(\mathbf{u})$ then $c'_{ij} = 0$. If (i, j) is any arc which is not in the tree, other than the profitable arc (p, q), the component of **v** along that arc is zero. Thus

$$\mathbf{c'v} = c'_{pq} v_{pq} \leqslant 0$$

Hence

$$\mathbf{cv} - \mathbf{cu} = \mathbf{c'v} - \mathbf{c'u} = \mathbf{c'v} = c'_{pq} v_{pq} \leqslant 0$$

Thus if we are able to locate a feasible tree solution **u** initially (an initialization procedure will be taken up in section 2.2) and if this FTS has a profitable arc, we can obtain another feasible tree solution **v** such that $\mathbf{cv} \leqslant \mathbf{cu}$. We continue this iteration process till we get an optimal feasible tree solution at which the algorithm terminates.

The cost of an updated FTS is less than the cost of the previous FTS if and only if $t > 0$. It may happen that $t = 0$, resulting in a new tree, but in this case the cost remains unchanged. This happens under degeneracy. In most problems degeneracy does not cause any trouble. But it is possible to construct examples in which degeneracy could cause the awkward phenomenon known as **cycling**: after several iterations we could come back to the same (degenerate) tree which we left earlier, never reaching an optimal feasible tree solution.

Before actually discussing an artificially constructed problem in which the rare phenomenon of cycling occurs, let us discuss in detail the solution of a simple problem which will clarify the ideas developed thus far in this section.

Example 2.1

Consider the network shown below with 7 vertices and 10 arcs in which any flow vector **x** is of the form

$$\mathbf{x}^T = [x_{13} \ x_{21} \ x_{32} \ x_{51} \ x_{52} \ x_{61} \ x_{64} \ x_{67} \ x_{74} \ x_{75}]$$

the cost vector is given by

$$\mathbf{c} = [1 \ 1 \ 1 \ 1 \ 5 \ 11 \ 20 \ 5 \ 5 \ 6]$$

and the supply–demand vector by

$$\mathbf{b}^T = [0 \ 2 \ 3 \ 5 \ -2 \ -3 \ -5]$$

(In the figure, S denotes supply and D denotes demand. Thus $S = 5$ at vertex 7 implies $b_7 = -5$. Likewise $b_4 = 5$.)

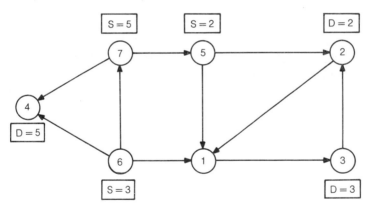

Iteration 1

We are given an initial feasible tree solution with which we start our first iteration.

Step 1. The initial FTS is

$$\mathbf{x}^T = [3 \ 0 \ 0 \ 0 \ 2 \ 3 \ 0 \ 0 \ 5 \ 0]$$

and its cost is $z = \mathbf{cx} = 71$. The arcs of a spanning tree T (not unique since **x** is degenerate) with which **x** can be associated are $(1, 3), (2, 1), (5, 2), (6, 1), (6, 7)$ and

$(7, 4)$. Using the defining relations, we obtain

$$\mathbf{y} = [6\ 5\ 7\ 5\ 0\ -5]$$

Step 2. We test whether there is a profitable arc e with respect to $\mathbf{x}$. In the present case we see that $e = (5, 1)$ is the only profitable arc which comes into $T(x)$ as an entering arc.
   Step 3. The cycle $C(e)$ is

$$
\begin{array}{cccc}
t & 0-t & 2-t \\
5 \longrightarrow 1 \longleftarrow 2 \longleftarrow 5
\end{array}
$$

in which $(5, 2)$ and $(2, 1)$ are backward arcs with existing flow 2 and 0, respectively. We revise the flow in this cycle as t units along $(5, 1)$, $2 - t$ units along $(5, 2)$ and $0 - t$ units along $(2, 1)$. Therefore $t = 0$. Since $(5, 1)$ has just entered it does not make sense to take it as a leaving arc. Thus $(2, 1)$ becomes the leaving arc. Since $t = 0$, the cost remains unchanged.

Iteration 2
Step 1. The current FTS is

$$\mathbf{x}^{T} = [3\ 0\ 0\ 0\ 2\ 3\ 0\ 0\ 5\ 0]$$

with cost $z = 71$. The arcs of the spanning tree are $(1, 3)$, $(5, 1)$, $(5, 2)$, $(6, 1)$, $(6, 7)$ and $(7, 4)$. From the defining equations,

$$\mathbf{y} = [6\ 10\ 7\ 5\ 5\ -5]$$

Step 2. We see that $e = (3, 2)$ is the only profitable arc which enters $T(\mathbf{x})$.
   Step 3. The cycle $C(e)$ is

$$
\begin{array}{cccc}
t & 2-t & 0+t & 3+t \\
3 \longrightarrow 2 \longleftarrow 5 \longrightarrow 1 \longrightarrow 3
\end{array}
$$

We get $t = 2$ and $(5, 2)$ leaves the tree.

Iteration 3
Step 1. Our FTS is now

$$\mathbf{x}^{T} = [5\ 0\ 2\ 2\ 0\ 3\ 0\ 0\ 5\ 0]$$

and $z = 67$. The spanning tree has arcs $(1, 3)$, $(3, 2)$, $(5, 1)$, $(6, 1)$, $(6, 7)$ and $(7, 4)$. Also

$$\mathbf{y} = [6\ 8\ 7\ 5\ 5\ -5]$$

Step 2. There are no profitable arcs for the current FTS. At this stage the algorithm halts after reporting that the current FTS is an optimal solution.

   Note that the initial FTS in this problem is degenerate, defining a spanning tree T in iteration 1 and another spanning tree T' in iteration 2, which resulted in two distinct dual solutions, one in each iteration. However, the existence of a degenerate FTS did not cause any complications and we successfully found an optimal FTS.

Degeneracy and cycling

As we remarked earlier, it is theoretically possible that degenerate feasible tree solutions can cause the extremely rare phenomenon of cycling, where we can go through iterations over and over again using the arcs of the same cycle (the cost being the same at the end of each iteration) without ever reaching an optimal solution. But the good news is that cycling can be avoided by a suitable choice of leaving arcs. For details regarding an elegant means of avoiding cycling, see Cunningham (1976).

We conclude this section with the presentation of a simplified version of a problem in which cycling can occur, as outlined in Chvátal (1980).

Example 2.2

In the set $V = \{1, 2, 3, 4, 5, 6, 7, 8\}$ of vertices of a network, the first four are sources with supply 1 at each source, and the next four vertices are sinks with demand 1 at each sink. The set of arcs is

$$A = \{(i, j): i \text{ is a source and } j \text{ is a sink}\}$$

The components of the cost vector $\mathbf{c} = [c_{ij}]$ are as follows:

$$c_{16} = c_{17} = c_{25} = c_{27} = c_{35} = c_{36} = c_{48} = 1$$

$c_{ij} = 0$ in all other cases. It is easy to see that the optimal cost is obviously $z = 0$.

Now consider an obvious feasible solution obtained when one unit is sent along each of the arcs $(1, 5)$, $(2, 6)$, $(3, 7)$ and $(4, 8)$ giving a total cost $z = 1$ and defining a degenerate FTS with an associated tree consisting of these four arcs and the arcs $(1, 6)$, $(2, 8)$ and $(4, 7)$.

If this feasible solution is taken as the initial FTS, at the first iteration the arc $(1, 8)$ can enter. The arc $(2, 8)$ leaves, giving another degenerate FTS for which also the cost is $z = 1$.

The process continues in this way as shown in Table 2.1, and we see that at the end of 12 iterations we have the tree with which we started. Notice that the FTS is the same for all the iterations but the associated trees vary from iteration to iteration till we come back to the same tree.

2.2 INITIALIZATION, FEASIBILITY AND DECOMPOSITION

There are only a finite number of spanning trees in a network, and so the network simplex algorithm should eventually terminate after a finite number of iterations, assuming that cycling does not take place. Specifically, this is the

Table 2.1 Cycling in a network: iterations performed in Example 2.1

Iteration number	Entering arc	Leaving arc	The other arcs in an associated tree
1	$(1, 8)$	$(2, 8)$	$(1, 5), (1, 6), (2, 6),$ $(4, 8), (4, 7), (3, 7)$
2	$(3, 6)$	$(1, 6)$	$(1, 8), (1, 5), (2, 6),$ $(4, 8), (4, 7), (3, 7)$
3	$(4, 6)$	$(4, 7)$	$(3, 6), (1, 8), (1, 5),$ $(2, 6), (4, 8), (3, 7)$
4	$(3, 5)$	$(3, 6)$	$(4, 6), (1, 8), (1, 5),$ $(2, 6), (4, 8), (3, 7)$
5	$(3, 8)$	$(1, 8)$	$(3, 5), (4, 6), (1, 5),$ $(2, 6), (4, 8), (3, 7)$
6	$(2, 5)$	$(3, 5)$	$(3, 8), (4, 6), (1, 5),$ $(2, 6), (4, 8), (3, 7)$
7	$(4, 5)$	$(4, 6)$	$(2, 5), (3, 8), (1, 5),$ $(2, 6), (4, 8), (3, 7)$
8	$(2, 7)$	$(2, 5)$	$(4, 5), (3, 8), (1, 5),$ $(2, 6), (4, 8), (3, 7)$
9	$(2, 8)$	$(3, 8)$	$(2, 7), (4, 5), (1, 5),$ $(2, 6), (4, 8), (3, 7)$
10	$(1, 7)$	$(2, 7)$	$(2, 8), (4, 5), (1, 5),$ $(2, 6), (4, 8), (3, 7)$
11	$(4, 7)$	$(4, 5)$	$(1, 7), (2, 8), (1, 5),$ $(2, 6), (4, 8), (3, 7)$
12	$(1, 6)$	$(1, 7)$	$(4, 7), (2, 8), (1, 5),$ $(2, 6), (4, 8), (3, 7)$

case if we make the assumption that every spanning tree appears at most once during the implementation of the algorithm. Thus termination of the algorithm is not an item of concern. The real problem, then, is to ascertain whether a given network has a feasible solution in the first place and to obtain an initial FTS if the problem is feasible.

Initialization and feasibility

Suppose there is a vertex v in the network such that: there is an arc from each source to v; there is an arc from v to each sink; and if w is an intermediate vertex then either (v, w) or (w, v) is an arc. Then it is easy to obtain an initial FTS. But a vertex satisfying these properties may not be readily located even if it exists.

In what follows we relax the assumption that the network is weakly connected. We choose an arbitrary vertex v and try to construct a tree T by constructing additional arcs as follows. If i is a source and if there is an arc from i to v then (i, v) is an arc of T. If there is no arc from i to v, we construct an arc from i to v and take this arc as an arc of T. Likewise, if j is a sink and if there is an arc from v to j, we take (v, j) as an arc of T and if there is no such arc we construct one and include it in the tree. Finally, if k is an intermediate vertex and if there is an arc from v to k or from k to v we take that arc as an arc of T and if there is none we construct an arc in either direction and take it as an arc of T.

The introduction of additional arcs (known as **artificial** arcs) as described above ensures that the enlarged network is feasible and weakly connected, with an easily discernible feasible tree solution. Obviously, if the given network is infeasible, every feasible solution in the larger network will have positive flow along at least one artificial arc. Let us agree to write a flow vector in the enlarged network G' (with r artificial arcs) as a vector with $m + r$ components, of which the first m components correspond to the m arcs of G. We define a new cost vector $\mathbf{c}'$ with $m + r$ components, of which the first m components are 0 and the remaining r components are 1.

Suppose $\mathbf{x}^*$ is an optimal feasible tree solution (with an associated tree T^*) of the transshipment problem (for the enlarged network G' with cost vector $\mathbf{c}'$ and the same demand vector $\mathbf{b}$ as in G), obtained by applying the network simplex method starting with an initial FTS (which is easily located by the way these arcs are constructed) in G'. There are three possibilities concerning the arcs of the tree T^* associated with the optimal solution $\mathbf{x}^*$ obtained using these artificial arcs. Let us examine these cases.

Case 1. No arc in T^* is artificial. In this case T^* is a tree in G which readily gives an initial FTS for the original network problem. Thus the original problem is feasible and can be easily initialized with this T^*.

Case 2. T^* has at least one artificial arc along which the flow is positive. So the optimal cost in G' is positive. Suppose the network G has a feasible solution $\mathbf{v}$. Let $\mathbf{v}'$ be the vector in which the first m components constitute the vector

v and the remaining components are all 0. Then **v'** is a feasible solution for the enlarged network such that $c'v' = 0$. This contradicts the optimality of **x***. So in this case the transshipment problem for G has no feasible solution.

Case 3. T^* has at least one artificial arc, but the flow in every artificial arc in T^* is zero. In this case the original problem has a feasible solution **x** obtained by the restriction of **x*** to the set of original arcs. There is an interesting difference between case 1 and this case. In the former we obtained a spanning tree in the original network and thus a feasible tree solution in it. In the present case, some of the arcs in T^* are artificial, and if we delete them we do not have a spanning tree. We have only an acyclic subgraph (with less than $n - 1$ arcs) of the network and thus the feasible solution is not an FTS. In this case the original problem decomposes into smaller feasible subproblems which may be solved separately.

Before we examine in detail the situation in which this kind of decomposition takes place, let us consider two examples – one infeasible and the other feasible – to clarify the first two cases.

Example 2.3

Show that the problem with $V = \{1, 2, 3, 4, 5, 6, 7\}$, flow vector

$$\mathbf{x} = [x_{12}\ x_{14}\ x_{23}\ x_{25}\ x_{35}\ x_{45}\ x_{46}\ x_{57}\ x_{67}]$$

and supply–demand vector

$$\mathbf{b} = [-6\ \ -4\ \ -8\ \ 0\ \ 0\ \ 11\ \ 7]$$

is infeasible.

We take vertex 5 as a fixed vertex and construct two artificial arcs $(1, 5)$ and $(5, 6)$. The cost vector for the enlarged problem is

$$\mathbf{c'} = [0\ 0\ 0\ 0\ 0\ 0\ 0\ 0\ 1\ 1]$$

Now we solve the problem for the enlarged network as follows:

Iteration 1 (initial FTS)

The arcs of the tree are $(2, 5)$, $(3,5)$, $(4, 5)$, $(5, 7)$ and the two artificial arcs $(1, 5)$ and $(5, 6)$. Then

$$\mathbf{x'} = [0\ 0\ 0\ 4\ 8\ 0\ 0\ 7\ 0\ 6\ 11]$$

The last two components in this vector correspond to the artificial arcs $(1, 5)$ and $(5, 6)$, respectively. Then

$$\mathbf{y} = [-1\ 0\ 0\ 0\ 0\ 1\ 0]$$

The arc $(1, 2)$ is profitable and it enters. Then the arc $(1, 5)$ leaves the tree, with $t = 6$.

Iteration 2
For the new tree, we obtain

$$\mathbf{y} = [0\ 0\ 0\ 0\ 0\ 1\ 0]$$

The arc $(4, 6)$ is profitable and it enters. Arc $(4, 5)$ leaves with $t = 0$.

Iteration 3
For the new tree,

$$\mathbf{y} = [0\ 0\ 0\ 1\ 0\ 1\ 0]$$

There are no profitable arcs. So the current solution is optimal. But the component of the optimal solution along the artificial arc $(5, 6)$ is not 0. So the given problem is infeasible.

Example 2.4
Consider the network obtained by replacing the arc $(4, 5)$ by the arc $(5, 4)$ in Example 2.3. Show that this modified problem is feasible. Obtain an initial FTS for this problem.
   The flow vector is

$$\mathbf{x} = [x_{12}\ x_{14}\ x_{23}\ x_{25}\ x_{35}\ x_{46}\ x_{54}\ x_{57}\ x_{67}]$$

We take vertex 5 to construct artificial arcs $(1, 5)$ and $(5, 6)$. The flow vector in the enlarged network is $\mathbf{x}'$, with x_{15} and x_{56} as its last two components. The first 9 components of $\mathbf{x}'$ constitute the flow vector $\mathbf{x}$. We now solve the problem for the extended network with the cost vector $\mathbf{c}'$, the first 9 components of which are 0 and the last two components are 1.

Iteration 1 (initial FTS)
The arcs in the tree are $(2, 5)$, $(3, 5)$, $(5, 4)$, $(5, 7)$ and the two artificial arcs. We obtain

$$\mathbf{y} = [-1\ 0\ 0\ 0\ 0\ 1\ 0]$$

The arc $(1, 2)$ is a profitable arc and it enters. The arc $(1, 5)$ leaves the tree.

Iteration 2
The arcs in the tree are $(1, 2)$, $(2, 5)$, $(3, 5)$, $(5, 4)$, $(5, 7)$ and the artificial arc $(5, 6)$. We obtain

$$\mathbf{y} = [0\ 0\ 0\ 0\ 1\ 0]$$

The only profitable arc is $(4, 6)$ which enters, and the remaining artificial arc $(5, 6)$ leaves.
   At this stage we have a spanning tree T^* in the enlarged network with no artificial arcs. So the problem is feasible. The feasible tree solution associated with this tree is the vector

$$[6\ 0\ 0\ 10\ 8\ 11\ 11\ 7\ 0].$$

Decomposition into subproblems

Suppose the set of vertices of the network is partitioned into two subsets W and W', and let the cardinality of W be k. If $\mathbf{b} = [b_i]$ is the supply–demand vector, the **net demand** $N(W)$ of W is the sum of the k numbers b_i, where $i \in W$.

If $\mathbf{x}$ is any feasible flow in the network, the **inflow** $I(W)$ into W is the sum of all x_{ij} where (i, j) is in E, $i \in W'$ and $j \in W$. Likewise, the **outflow** $O(W)$ from W is the sum of all x_{ij} where (i, j) is an arc of the network, $i \in W$ and $j \in W'$.

Without loss of generality we can assume that the first k rows of the incidence matrix $\mathbf{A}$ of the network correspond to the k vertices of the set W. Let $\mathbf{A}'$ be the $k \times m$ submatrix formed by these k rows and the columns of $\mathbf{A}$ and let $\mathbf{b}'$ be the subvector of $\mathbf{b}$ consisting of the first k components. If we add together the k linear equations in the linear system $\mathbf{A}'\mathbf{x} = \mathbf{b}'$ we find that the sum on the left-hand side is $I(W) - O(W)$ and the sum on the right-hand side is $N(W)$. Thus $I(W) - O(W) = N(W)$ for any set W of vertices. In other words, the net demand is always inflow minus outflow. This relation confirms the intuitive feeling that the net demand is 'what comes in minus what goes out'.

A set W of vertices in the network is called an **autonomous set** if its net demand $N(W)$ is 0 and if there are no arcs in the network from vertices in the complement of W to vertices in W. It is easy to see that if W is an autonomous set and $\mathbf{x}$ is a feasible flow, then $x_{ij} = 0$ for every arc from a vertex in W to a vertex not in W. Thus an autonomous set is a set of vertices which is 'self-sufficient' as a collection of vertices, has no arcs to import the commodity from outside, and will not export the commodity to points outside.

Thus if we can identify an autonomous set W in a feasible network we can decompose the original problem into two smaller feasible subproblems, one involving W and the other involving its complement W' by ignoring all arcs from vertices in W to vertices in W'. In the opposite direction, we prove in Theorem 2.4 that when the problem decomposes into feasible subproblems as in case 3 above, we can always locate an autonomous set in the network.

Theorem 2.4
Let $V = \{1, 2, \ldots, n\}$ be the set of vertices of the feasible network. Let (i, j) be an artificial arc in the optimal tree T^* associated with an optimal solution $\mathbf{x}^*$ of the enlarged problem, and let $\mathbf{y} = [y_1 \ y_2 \cdots y_n]$ be the corresponding dual solution. If $W = \{k : y_k \leqslant y_i\}$, then the complement of W is an autonomous set.

Proof
By definition i is in W. So W is nonempty. Also $y_i + 1 = y_j$, which implies j is in W'. Thus W' is a proper subset of V.

Suppose there is an arc (p, q) in the original network, where p is in W and q is in W'. Then $y_p \leqslant y_i < y_q$, which implies $y_p < y_q$. But no arc is profitable as far as T^* is concerned in the extended network, which implies $y_q \leqslant y_p$. So in the original network there are no arcs from a vertex in W to a vertex in W'. It remains to be shown that $N(W') = 0$.

Suppose $\mathbf{x}^*$ is the optimal flow in the enlarged network associated with T^*. Then $N(W') = I(W') - O(W')$.

The component of $\mathbf{x}^*$ along an arc which is not in T^* is zero. So to compute the left-hand side we have to consider only those arcs which are in T^* and which are not artificial since the original problem is feasible. Thus $I(W') = 0$.

Suppose there is an original arc (a, b) which is also an arc in T^* where a is in W' and b is in W. Then we have a contradiction since $y_b < y_a$ and $y_a + 0 = y_b$. Thus (a, b) is not in T^*, which implies the component of $\mathbf{x}^*$ along the arc (a, b) is 0. Hence $O(W')$ is also 0. So $N(W') = 0$. This completes the proof.

A simple example to illustrate the property established in the theorem is now given.

Example 2.5
Consider the network G with vertex set $V = \{1, 2, 3, 4\}$ and arc set $E = \{(1, 2),$ $(2, 3), (4, 3), (1, 4)\}$. The flow vector is $\mathbf{x} = [x_{12} \ x_{23} \ x_{43} \ x_{14}]$, and the supply–demand vector is $\mathbf{b} = [-4 \ -6 \ 6 \ 4]$. It is easy to see that $\{1, 4\}$ is an autonomous set.

Let us enlarge the network by keeping vertex 1 as the starting vertex. Then the artificial arcs are $(2, 4)$ and $(1, 3)$. An optimal feasible tree solution in the larger problem is $[0 \ 6 \ 0 \ 4 \ 0 \ 0]$.

We can take $(1, 4), (2, 3)$ and $(2, 4)$ as the arcs of an optimal tree in which the flow along the artificial arc $(2, 4)$ is zero. The dual solution corresponding to this tree is $[0 \ -1 \ -1 \ 0]$. Then $W = \{2, 3\}$ and $W' = \{1, 4\}$.

We can summarize the foregoing in the form of an algorithm:

Step 1. Given the directed network $G = (V, E)$, construct the enlarged network $G' = (V, E')$. The set of artificial arcs is $A' = E' - E$.

Step 2. Solve the problem for G' with the cost vector $\mathbf{c}'$. Let $\mathbf{x}^*$ be an optimal tree solution of G' with T^* as the spanning tree.

Step 3. Does T^* have an artificial arc? If the answer is no, then T^* is a spanning tree in G, giving an initial FTS for the given problem. Use this FTS to initialize the problem and obtain an optimal solution of the problem. Go to step 6. If the answer is yes, then go to step 4.

Step 4. Is there an artificial arc in T^* such that the component of $\mathbf{x}^*$ along that arc is positive? If the answer is yes, the problem is infeasible. Go to step 6. Otherwise go to step 5.

Step 5. At this stage we know that T^* has at least one artificial arc and the component of $\mathbf{x}^*$ along every artificial arc is 0. The problem is feasible and decomposes into feasible subproblems. Obtain an FTS for each of them by repeating the process if necessary. Go to step 6.

Step 6. Stop.

Integrality of flows and unimodularity

In many cases, a feasible solution of the optimization problem is relevant only when each component of the solution is a nonnegative integer. Consider, for example, the case where the supply at a source is the number of trucks manufactured there and the demand at a sink is the number of trucks needed there in a single-commodity problem involving the shipment of trucks. In such situations we are interested in the 'integrality' of the optimal solution.

If a feasible nonoptimal tree solution in the transshipment problem has integer components and if it has a profitable arc e, then in the iteration using e as the entering arc, the flow along e is an integer and therefore the components of the feasible tree solution in the next iteration also are integers. Thus if the initial FTS has integer components we are assured of the existence of an optimal solution in integers.

A linear programming (LP) problem which seeks an optimal solution in which each component is an integer is called an **integer linear programming** (ILP) problem. The LP problem obtained by removing the restriction of integrality in an ILP problem is called the **LP relaxation** of the ILP problem. If an optimal solution $\mathbf{x}$ of the LP relaxation has only integer components, then $\mathbf{x}$ must be an optimal solution of the ILP problem as well. However, if $\mathbf{x}$ has noninteger components and if we approximate $\mathbf{x}$ by a feasible $\mathbf{x}'$ with integer components, it is not all necessary that $\mathbf{x}'$ will be an optimal solution of the ILP problem. But as we remarked earlier, in the case of a feasible transshipment problem (in which the supply–demand vector has only integer components) if we relax the integrality requirement and solve the problem by the network simplex method with an initial FTS with integer components, we obtain an optimal solution with integer components after a finite number of iterations.

This built-in integrality property of a feasible solution of the transshipment problem is a consequence of the unimodularity of the incidence matrix of the network. Notice that if $\mathbf{A}'$ is a nonsingular submatrix of a totally unimodular matrix $\mathbf{A}$, then the unique solution of the linear system $\mathbf{A}'\mathbf{x} = \mathbf{b}$, where $\mathbf{b}$ is an integer vector, is itself an integer vector since each component of this solution is of the form (p/q), where p is an integer (being the determinant of a matrix in which each entry is 0 or 1 or -1) and $q = \det \mathbf{A}'$, which is either 1 or -1 since $\mathbf{A}$ is unimodular.

2.3 INEQUALITY CONSTRAINTS

So far we have considered only those problems where the total supply and total demand are equal. If the total supply exceeds the total demand, with an excess supply of r units of the commodity, we construct a dummy vertex v as a sink vertex with demand r and construct arcs from each source vertex to the dummy vertex with the stipulation that the component of the extended cost vector is 0 along each of these artificial arcs. In other words, we dump the excess supply at no additional cost and solve the optimization problem at the same time.

If the total demand exceeds the total supply, creating an excess demand of r units, the problem in a practical sense is infeasible. However, we can construct a dummy vertex from which we hope to get a supply (wishful thinking!) of the extra r units needed. In this case we construct artificial arcs from the dummy vertex to each sink vertex. This time the stipulation is that the cost of sending one unit of the commodity along any of these artificial arcs is prohibitively high.

Specifically, if the total supply exceeds the total demand, then the linear equality constraint at each source vertex i will be replaced by the requirement that the inflow at i minus the outflow at i is greater than or equal to $-b_i$. If the total demand exceeds the total supply, then the constraint at each sink vertex j will be replaced by the requirement that the inflow at i minus the outflow at i is less than or equal to b_j. So the linear system of constraints can be written as $\mathbf{Ax} \geqslant \mathbf{b}$ or $\mathbf{Ax} \leqslant \mathbf{b}$, where $\mathbf{A}$ is the incidence matrix.

Hence the transshipment problem in which the equilibrium condition does not hold is a special case of the integer linear programming problem: Given an $n \times m$ constraint matrix $\mathbf{A}$, a $1 \times m$ cost vector $\mathbf{c}$ and an $n \times 1$ vector $\mathbf{b}$, find a vector $\mathbf{x}$ in which each component is a nonnegative integer such that $\mathbf{Ax} \leqslant \mathbf{b}$ and $\mathbf{cx}$ is as small as possible.

If it is possible to convert a general ILP problem of this type into a transshipment problem such that the constraint $\mathbf{Ax} \leqslant \mathbf{b}$ is replaced by $\mathbf{A'x'} = \mathbf{b'}$, where $\mathbf{A'}$ is the incidence matrix of a directed network, then we can solve the ILP by solving the associated transshipment problem.

The procedure is as follows:

Step 1. Convert each inequality constraint into a $\leqslant$ inequality constraint and add a nonnegative **slack variable** to that constraint to make it into an equality constraint.

Step 2. Introduce a linear equality constraint on the slack variables: the slack variables should sum to r, where the value of r is to be determined.

Step 3. Let the linear system thus obtained be $\mathbf{A'x'} = \mathbf{b'}$. If there is a digraph G such that $\mathbf{A'}$ is its incidence matrix (by changing the signs of the coefficients in $\mathbf{A'}$ and $\mathbf{b'}$ if necessary), then the problem can be solved by the network simplex method.

Step 4. Find r such that the sum of all the components of $\mathbf{b'}$ is 0. We now have a network in which any arc corresponding to a slack variable is an artificial arc.

Step 5. Let $\mathbf{c'}$ be the vector such that (i) a component corresponding to an artificial arc is 0, and (ii) $\mathbf{c}$ is a subvector of $\mathbf{c'}$.

Step 6. Solve the transshipment problem with $\mathbf{A'}$, $\mathbf{b'}$ and $\mathbf{c'}$. If $\mathbf{x'}$ is an optimal solution of this problem, then the subvector $\mathbf{x}$ is an optimal solution of the ILP problem.

Here is an illustrative example.

Example 2.6

The cost vector is $\mathbf{c} = [4\ 3\ 9\ 2\ 7]$ and the constraints are

$$
\begin{array}{rrrrrcr}
1 & 1 & 1 & 0 & 0 & = & 5 \\
-1 & 0 & 0 & 1 & 0 & \leqslant & 0 \\
0 & 1 & 0 & 0 & -1 & = & -2 \\
0 & 0 & 1 & 1 & 1 & \geqslant & 3
\end{array}
$$

We introduce two slack variables, and the system becomes

$$
\begin{array}{rrrrrrrcr}
1 & 1 & 1 & 0 & 0 & 0 & 0 & = & 5 \\
-1 & 0 & 0 & 1 & 0 & 1 & 0 & = & 0 \\
0 & 1 & 0 & 0 & -1 & 0 & 0 & = & -2 \\
0 & 0 & -1 & -1 & -1 & 0 & 1 & = & -3 \\
0 & 0 & 0 & 0 & 0 & 1 & 1 & = & r
\end{array}
$$

We can then rewrite the above system as

$$
\begin{array}{rrrrrrrcr}
-1 & -1 & -1 & 0 & 0 & 0 & 0 & = & -5 \\
1 & 0 & 0 & -1 & 0 & -1 & 0 & = & 0 \\
0 & 1 & 0 & 0 & -1 & 0 & 0 & = & -2 \\
0 & 0 & 1 & 1 & 1 & 0 & -1 & = & 3 \\
0 & 0 & 0 & 0 & 0 & 1 & 1 & = & 4
\end{array}
$$

This system is $\mathbf{A}'\mathbf{x}' = \mathbf{b}'$. We now solve the transshipment problem with $\mathbf{A}'$, $\mathbf{b}'$ and $\mathbf{c}'$ where $\mathbf{c}' = [4\ 3\ 9\ 2\ 7\ 0\ 0]$. An optimal solution for the transshipment problem is $\mathbf{x}' = [5\ 0\ 0\ 1\ 2\ 4\ 0]^{\mathrm{T}}$. Thus an optimal solution for the ILP problem is $\mathbf{x} = [5\ 0\ 0\ 1\ 2]^{\mathrm{T}}$.

2.4 TRANSPORTATION PROBLEMS

A transshipment problem is called a **transportation problem** if there are no intermediate vertices in the network and if every arc in it is from a source to a sink. Thus the set of vertices of the network can be partitioned into two subsets S (the set of sources) and D (the set of sinks). The underlying graph is then a bipartite graph $G = (S, D, E)$. In our discussion of the transportation problem we will see that once the procedure is made clear it really does not matter whether the network is directed or not. We also recall that the incidence matrix of a digraph and the incidence matrix of a bipartite graph are both totally unimodular, assuring the integrality of the flow if the supply–demand vector is an integer vector. If there is an arc from every source to every sink (or in other words the underlying bipartite graph is complete) the problem is known as a **Hitchcock problem** or a **Hitchcock–Koopmans problem**. If there is no arc from a source to a sink in a transportation problem, we assume that there is one but that its use for transporting the commodity is prohibitively expensive. Thus any transportation problem can be considered as a Hitchcock problem.

In a Hitchcock problem if S_i $(i = 1, 2, \ldots, m)$ are the sources and D_j $(j = m + 1, m + 2, \ldots, m + n)$ are the sinks, the flow vector with mn components can also be represented as an $m \times n$ matrix $[x_{ij}]$ where x_{ij} is the flow from S_i to D_j. So each feasible flow can be viewed as a nonnegative matrix known as the **flow matrix**. For notational convenience, from now on we use the subscript i to denote the source S_i and the subscript j to denote the sink D_j. Thus the arc (i, j) is the arc from the source S_i to the sink D_j. The **cost matrix c** is an $m \times n$ matrix $[c_{ij}]$, c_{ij} being the cost of sending one unit of the commodity from the source i to the sink j. The total cost is the inner product $\mathbf{c} \cdot \mathbf{x}$ which is the sum of the mn pairwise numbers $c_{ij} x_{ij}$. If the problem is not a Hitchcock problem, then there will be at least one i and one j such that (i, j) is not an arc of the network, in which case we let $c_{ij} = M$, where M is a large positive number. A solution is obviously feasible if and only if it does not have a component x_{ij} where $c_{ij} = M$.

The transportation tableau

The solution to a transportation problem can be represented in a tableau in which each row represents a source and each column represents a sink. The element in the (i, j)th cell in the tableau is the amount x_{ij} of flow from source i to sink j. The sum of the elements in a row is the supply at the source corresponding to that row. Likewise, the sum of the elements in a column is the demand at the vertex corresponding to that column. The cost c_{ij} is entered in the (i, j)th cell at the top left-hand corner of the cell. This tabular representation does not make it easier to solve the problem, but makes the iteration process in the manual solution of a problem with a small number of vertices more convenient.

For example, a feasible solution (not an FTS) of a transportation problem with three sources and four sinks in which total supply and total demand are both equal to 100 is represented in tableau form below. It is not a Hitchcock

	4	5	6	7	Supply
1	8 15	6 15	3	M	30
2	6 5	4 10	7 15	M	30
3	M	3 5	9 15	8 20	40
Demand	20	30	30	20	100

problem. From source 1, we send 15 units to sink 4 and 15 units to sink 5. From source 2, we send 5 units to sink 4, 10 units to sink 5 and 15 units to sink 6. Finally, from source 3, we send 5 units to sink 5, 15 units to sink 6 and 20 units to sink 7. The total cost for this feasible flow is

$$15 \times 8 + 15 \times 6 + 5 \times 6 + 10 \times 4 + 15 \times 7 + 5 \times 3 + 15 \times 9 + 20 \times 8 = 695$$

Finding an initial FTS

Consider a Hitchcock problem in which the supply at S_i is s_i for each i in $\{1, 2, \ldots, m\}$ and the demand at D_j is d_j for each j in $\{m+1, m+2, \ldots, m+n\}$. Assume that the total supply s and the total demand d are equal. An initial FTS can easily be obtained by the following procedure.

Start from S_1. Exhaust all the supply available there by sending flow first to D_1. If all the needs of D_1 are met then send flow from S_1 to D_2. On the other hand, if S_1 cannot meet the needs of D_1, go to S_2 and try to meet the remaining needs of D_1 by using the supply available at S_2. Continue this process till all the available supply is exhausted. This method, known as the **northwest corner method**, gives an FTS in which some of the components may be 0. The number of 'used' cells must be exactly $m+n-1$, which is precisely the number of arcs in the tree.

An initial FTS obtained by this method for the data given in the above tableau is as follows:

	4	5	6	7	Supply	y_i
1	8 (20)	6 −(10)	3 −6 +	M	30	−8
2	6 +	4 +(20)	7 (10) −	M	30	−6
3	M	3	9 (20)	8 (20)	40	−8
Demand	20	30	30	20	100	
y_j	0	−2	1	0		

The total cost of this feasible flow is

$$20 \times 8 + 10 \times 6 + 20 \times 4 + 10 \times 7 + 20 \times 9 + 20 + 8 = 710$$

Is this an optimal flow? To ascertain this we conduct an optimality test. If this solution is not optimal, we iterate and move to a 'better' feasible solution.

Moving from a nonoptimal tableau to the next tableau

As we move from one iteration to the next, we move from one tableau to the next.

We start with a feasible tree as usual and compute the unique dual solution **y** with $m + n$ components in which the last component is 0. For each source, we have a number y_i $(i = 1, 2, \ldots, m)$ and for each sink we have a number y_j $(j = m + 1, m + 2, \ldots, m + n)$ which satisfy the defining relation $y_i + c_{ij} = y_j$ whenever (i, j) is an arc in the tree.

The numbers y_i $(i = 1, 2, \ldots, m)$ are written as a column on the right-hand side of the tableau and the numbers y_j are written as a row at the bottom of the tableau (see the FTS tableau above). The last component of y_j, when $j = m + n$, is taken as 0.

If (i, j) is an arc of the tree then enter the nonnegative number x_{ij} in that cell and circle that. Thus there will be exactly $m + n$ 1 circled entries in any feasible tree tableau. In each uncircled cell where the cost is not M, we compute the number $d_{ij} = c_{ij} + y_i - y_j$. The current flow is optimal if and only if d_{ij} is nonnegative in each uncircled cell. Otherwise, choose an uncircled cell with a negative value of d_{ij}; the arc corresponding to this cell is an entering arc.

In the FTS tableau above, $d_{16} = c_{16} + y_1 - y_6 = 3 + (-8) - 1 < 0$. So this solution is not optimal and the arc $(1, 6)$ enters the tree.

Once the entering cell is selected we look for a leaving cell. If we adjoin an entering arc e to the tree, we have a unique cycle $C(e)$ in which an arc is forward if it has the same direction as e, and is backward otherwise. The underlying graph is bipartite. So every cycle has an even number of arcs. If the arcs of the cycle $C(e)$ are arranged sequentially starting from e, they will be alternately forward and backward.

In our example the arc $(1, 6)$ enters the tree, creating the cycle $C(e)$ which has the following four arcs in sequence: $(1, 6), (2, 6), (2, 5)$ and $(1, 5)$. Here $(1, 6)$ and $(2, 5)$ are forward and the other two are backward. When we move from the forward arc $(1, 6)$ to the backward arc $(2, 6)$, in the tableau we are moving along a column from the entering uncircled cell $(1, 6)$ to the next circled cell $(2, 6)$. The next move from the backward arc $(2, 6)$ to the forward arc $(2, 5)$ represents a movement along a row in the tableau from the cell $(2, 6)$ to the cell $(2, 5)$.

Thus a move in the cycle from a forward arc to a backward arc involves a move in the tableau along a column, and a move in the cycle from a backward arc to a forward arc involves a move along a row in the tableau. These moves take place alternately, coming back ultimately to the uncircled cell passing through an odd number of circled cells. In the tableau the cells representing the arcs of the cycle are linked together by vertical and horizontal lines without showing their directions explicitly, as in the tableau on page 58.

Once the entering cell is selected, we look for a leaving cell. We add t units of flow along the entering arc and t more units of flow along each forward arc. At the same time we subtract t units of flow from each backward arc, making sure that the flow in no backward arc is negative. An arc in the cycle in which the flow becomes zero is selected as the leaving arc.

The following streamlined procedure to locate a leaving cell in the tableau is nothing but a reformulation of the iteration process established during the course of the proof of the network simplex algorithm.

First, we put a square in the entering cell and start from that cell. We can move either in a vertical or a horizontal direction from cell to cell. When we reach a used cell, we either ignore that cell and continue along the same direction or include it in the cycle and change direction from vertical to horizontal or from horizontal to vertical. Since the cycle $C(e)$ is unique, we come back to the entering cell after traversing all the arcs of the cycle. The entering cell is marked with a plus sign and others in the cycle are alternately marked plus and minus. The number of cells marked plus will of course be the same as the number of cells marked minus. We add t units to the entering cell. If p is the number in a circled entry in the cycle, then we make it $p + t$ if the cell is marked plus or $p - t$ if it is marked minus. Now we choose t such that the number in at least one circled entry becomes 0 and no circled entry is negative. We choose a cell (different from the entering cell) which gives a circled entry with 0. This cell is then the leaving cell. We now have an FTS whose cost does not exceed the cost of the FTS with which we started.

In our example, we put t units in the entering cell $(1, 6)$. Then we put $10 - t$ units in $(2, 6)$, $20 + t$ units in $(2, 5)$ and finally $10 - t$ units in $(1, 5)$. Thus $t = 10$ and either $(1, 5)$ or $(2, 6)$ can leave the tree. If we decide that $(2, 6)$ leaves the tree, the arcs in the next FTS are $(1, 4), (1, 5), (1, 6), (2, 5), (3, 6)$ and $(3, 7)$. The updated flow is as depicted below.

	4	5	6	7	Supply	y_i
1	8 20	6 0	3 10	M	30	
2	6	4 30	7	M	30	
3	M	3	9 20	8 20	40	
Demand	20	30	30	20	100	
y_j						

The total cost of the updated flow is

$$20\times 8+0\times 6+10\times 3+30\times 4+20\times 9+20\times 8=650$$

which is less than the total cost of the previous FTS, as it should be.

Example 2.7
Consider a problem with three sources and five sinks as in the tableau below. Use the procedure described above to locate a leaving cell if the entering cell is (3, 7). (Notice that once the entering cell is fixed, the cost matrix is not used in locating a leaving cell.)

	4	5	6	7	8	Supply
1	− (35)		(20)		(10) +	65
2	+ (5)	(20)		(30) −		55
3				+ □	(50) −	50
Demand	40	20	20	30	60	170

We start from the uncircled cell (3, 7). We could move vertically or horizontally. We proceed horizontally to the circled cell (3, 8), then vertically to (1, 8) and then horizontally to (1, 4) after ignoring the circled entry (1, 6). If we do not skip the (1, 6) and decide to include it, we will have to change our direction at (1, 6) and we will be stuck. From (1, 4) we move to (2, 4) and then to (2, 7). Finally, we move from (2, 7) to the uncircled cell (3, 8) which was the starting point. It is easy to see that $t=30$ and (2, 7) is the leaving cell.

2.5 SOME APPLICATIONS TO COMBINATORICS

A **matching** in a graph is set M of edges such that no two edges in M have a vertex in common. A matching M in a bipartite graph $G=(X, Y, E)$, where both X and Y have n vertices, is called a **perfect matching** if there are n edges in M. We will consider optimization problems involving matchings in Chapter 5. In this section we will see some applications of the theory of flows in transshipment networks to combinatorics.

A graph is called a **regular** graph of degree k if the degree of each vertex is k. As an immediate consequence of the integrality property established in section 2.2, we have the following theorem.

***Theorem 2.5* (Konig's marriage theorem)**
A regular bipartite graph of degree k (where $k>0$) has a perfect matching.

Proof

Let $G = (X, Y, E)$ be a regular bipartite graph of degree k where k is a positive integer and let the number of vertices in both X and Y be n.

Consider each vertex in X as a source, each vertex in Y as a sink and each edge $\{x, y\}$ as an arc (x, y) from a vertex x in X to a vertex y in Y. Suppose the supply at each source is 1 and the demand at each sink is also 1. We now have a transportation problem which has a feasible solution the component of which along each arc is $1/k$. So it has an optimal solution with integer components since the supply–demand vector is an integer vector. The n edges with positive flow corresponding to an optimal solution constitute a perfect matching in the graph.

(The classical result (dating from 1916) proved above is due to Konig and is often stated in matrimonial parlance as follows. Suppose that, in a party of unmarried people, each woman has been previously acquainted with exactly k men in the party and each man has been previously acquainted with exactly k women in the party. Then the women and the men in the party can be paired such that the woman and the man in each pair have been mutual acquaintances before coming to the party. Acquaintance, of course, is a symmetric relation: if x is acquainted with y then y is acquainted with x. In this book, this theorem is called Konig's marriage theorem to distinguish it from Theorem 2.8, which is known as Konig's theorem.)

Notice that in the course of proving Theorem 2.5, the 'level' of acquaintance between two individuals was not taken into account. If A is acquainted with both B and C, it is not necessary in general that A's levels of acquaintance with B and C are the same. So it is interesting to see whether the theorem is true in a more general setting.

First, we notice that the $n \times n$ initial flow matrix introduced during the proof of the theorem has a special property: all row sums and all column sums of the flow matrix are equal to 1. A nonnegative square matrix with this property is called a **doubly stochastic matrix**. A doubly stochastic integer matrix $\mathbf{P}$ is a permutation of the identity matrix $\mathbf{I}$ such that $\mathbf{PP}^T = \mathbf{I}$.

Given a doubly stochastic $n \times n$ matrix we can define a transportation problem with n sources (with supply equal to 1 at each source) and n sinks (with demand equal to 1 at each sink) in which an arc (i, j) exists in the bipartite graph if and only if the (i, j)th element in the matrix is positive. This problem has a feasible solution, the component of which along the arc (i, j) is the (i, j)th element in the matrix. So it has a feasible solution with integer components. In other words, we have the following generalization of the theorem: for each doubly stochastic $n \times n$ matrix $\mathbf{D} = [d_{ij}]$, there corresponds a permutation $\mathbf{P} = [p_{ij}]$ of the $n \times n$ identity matrix such that $d_{ij} = 0$ implies $p_{ij} = 0$.

(According to Lawler (1976), this generalization has been cited as 'proof' that monogamy is the best of all possible systems of marriage:

Suppose we have a society of n women and n men. Let c_{ij} represent the benefit to be derived from full time cohabitation of woman i with man

j and let x_{ij} represent the fraction of time that i actually cohabits with j. If the objective is to maximize the total benefit, so the argument goes, there is an optimal solution in which x_{ij} is either 0 or 1.

This corresponds to a situation in which cohabitation is restricted to one person only.)

Theorem 2.6
If $\mathbf{D}$ is a doubly stochastic $n \times n$ matrix which has fractional elements, then there exist positive numbers a and b such that $\mathbf{D} = a\mathbf{P} + b\mathbf{D}'$, where $a + b = 1$, $\mathbf{P}$ is a permutation of the $n \times n$ identity matrix and $\mathbf{D}'$ is a doubly stochastic matrix which has more zeros than $\mathbf{D}$.

Proof
Since $\mathbf{D} = [d_{ij}]$ has fractional elements, it is not a permutation of the identity matrix. So there should be a permutation $\mathbf{P}$ of the identity such that $d_{ij} = 0$ implies $p_{ij} = 0$ due to the integrality theorem. Let a be the smallest d_{ij} such that p_{ij} is positive. Since $\mathbf{D}$ is not a permutation matrix, a is less than 1. Then we can write $\mathbf{D}$ as the sum $a\mathbf{P} + (1 - a)\mathbf{D}'$, where $\mathbf{D}'$ is doubly stochastic with more zero elements than $\mathbf{D}$. This completes the proof.

If a_i $(i = 1, 2, \ldots, k)$ are nonnegative numbers whose sum is 1, then $a_1\mathbf{M}_1 + a_2\mathbf{M}_2 + \cdots + a_k\mathbf{M}_k$ is a **convex combination** of the $n \times n$ matrices $\mathbf{M}_i$ $(i = 1, 2, \ldots, k)$. The following theorem characterizes doubly stochastic matrices.

Theorem 2.7 (Birkhoff–von Neumann theorem)
A matrix is doubly stochastic if and only if it is a convex combination of permutation matrices.

Proof
Obviously, any convex combination of permutation matrices is a doubly stochastic matrix. To prove the result in the opposite direction, we adopt a constructive procedure based on Theorem 2.6.

First, write the doubly stochastic matrix $\mathbf{D}$ as the convex combination $\mathbf{D} = a_1\mathbf{P}_1 + b_1\mathbf{D}_1$, where $\mathbf{D}_1$ has more zeros than $\mathbf{D}$. Then write $\mathbf{D}_1$ as the convex combination $\mathbf{D}_1 = a_{12}\mathbf{P}_2 + b_{12}\mathbf{D}_2$, where $\mathbf{P}_2$ is a permutation matrix and $\mathbf{D}_2$ is doubly stochastic with more zeros than $\mathbf{D}_1$. Let $a_2 = b_1 a_{12}$ and $b_2 = b_1 b_{12}$. Then $\mathbf{D} = a_1\mathbf{P}_1 + a_2\mathbf{P}_2 + b_2\mathbf{D}_2$ is a convex combination.

Continue this process till all the matrices in the convex combination are permutation matrices. Since each iteration produces more zeros in the last matrix on the left-hand side, the procedure must eventually terminate.

Example 2.8
Express the 3×3 doubly stochastic matrix $\mathbf{D}$ as the convex combination of permutation matrices.

$$\mathbf{D} = \begin{bmatrix} \frac{1}{2} & \frac{1}{2} & 0 \\ \frac{1}{4} & 0 & \frac{3}{4} \\ \frac{1}{4} & \frac{1}{2} & \frac{1}{4} \end{bmatrix}$$

A permutation matrix $\mathbf{P}$ which corresponds to $\mathbf{D}$ such that the zero elements match with the zero elements in $\mathbf{P}$ is

$$\mathbf{P} = \begin{bmatrix} 0 & 1 & 0 \\ 1 & 0 & 0 \\ 0 & 0 & 1 \end{bmatrix}$$

The smallest number in $\mathbf{D}$ which corresponds to a positive element in $\mathbf{P}$ is $\frac{1}{4}$. So we write $\mathbf{D} = \frac{1}{4}\mathbf{P} + \frac{3}{4}\mathbf{D}'$, where

$$\mathbf{D}' = \begin{bmatrix} \frac{2}{3} & \frac{1}{3} & 0 \\ 0 & 0 & 1 \\ \frac{1}{3} & \frac{2}{3} & 0 \end{bmatrix}$$

Now we write $\mathbf{D}'$ as the convex combination $\frac{1}{3}\mathbf{P}' + \frac{2}{3}\mathbf{D}''$, where

$$\mathbf{P}' = \begin{bmatrix} 0 & 1 & 0 \\ 0 & 0 & 1 \\ 1 & 0 & 0 \end{bmatrix} \quad \text{and} \quad \mathbf{D}'' = \begin{bmatrix} 1 & 0 & 0 \\ 0 & 0 & 1 \\ 0 & 1 & 0 \end{bmatrix}$$

At this stage $\mathbf{D}''$ is also a permutation of the identity matrix and so we stop. Thus a convex combination of the 3×3 doubly stochastic matrix $\mathbf{D}$ is $\frac{1}{4}\mathbf{P}_1 + \frac{1}{4}\mathbf{P}_2 + \frac{1}{2}\mathbf{P}_3$ where $\mathbf{P}_1 = \mathbf{P}$, $\mathbf{P}_2 = \mathbf{P}'$ and $\mathbf{P}_3 = \mathbf{D}''$ are all permutations of the identity matrix.

Matchings and coverings

A **covering** in a graph is a set W of vertices such that every edge of the graph is incident to at least one of the vertices in W.

The relation between matchings and coverings is not immediately apparent. Suppose M is an arbitrary matching consisting of k edges. Then M will be joining $2k$ distinct vertices pairwise. So any covering should have at least k vertices. Thus the cardinality of a matching cannot exceed the cardinality of a covering and consequently the size of a maximum matching is always less than or equal to the size of a minimum covering in a graph. In the case of a bipartite graph the surprising fact is that these two numbers are equal, as shown in Theorem 2.8.

Before proving this, let us have a closer look at matchings and coverings in bipartite graphs in the context of a transshipment problem.

Suppose $G = (X, Y, E)$ is a bipartite graph with m vertices in X and n vertices in Y. X is the set of left vertices and Y is the set of right vertices. Construct a digraph G' as follows. Each edge between a vertex p in X and a vertex q in

Y becomes an arc (p, q) from the left vertex p to the right vertex q. Construct a vertex v and draw arcs from each left vertex to v. Construct a vertex w and draw arcs from w to each right vertex in Y. Finally construct an arc from w to v (see graph B in Example 2.9).

We now define a transshipment problem on the digraph G' thus constructed as follows. The m left vertices and vertex w are sources. The other vertices are all sinks. The supply at each left vertex is 1 and the supply at w is n. The demand at each right vertex is 1 and the demand at v is m. Thus total supply and total demand are both equal to $m + n$.

Any feasible integer-valued solution of this transshipment problem defines a unique matching in the bipartite graph G. This is because the flow along any original arc is either 0 or 1 and the arcs with flow equal to 1 constitute a matching. The component along (w, v) of the integer-valued feasible solution $\mathbf{x}$ is denoted by x_{wv}. Now exactly $n - x_{wv}$ vertices in Y have their demands met from w. The remaining x_{wv} vertices in Y have their demands met from vertices in X. Thus any feasible integer-valued solution of the transshipment problem defines a matching in G consisting of x_{wv} arcs. Conversely, every matching involving k edges defines a unique integer-valued solution $\mathbf{x}$ of the transshipment problem such that $x_{wv} = k$.

Hence a largest size matching in the bipartite graph G can be obtained by solving the transshipment problem 'minimize $z = cx$' in this digraph G' using the network simplex method, where all the components of the cost vector $\mathbf{c}$ are zero except the component c_{wv} which is taken as -1.

The dual solution $\mathbf{y}$ corresponding to an FTS will have $m + n + 2$ components. Each component of $\mathbf{y}$ corresponds to a vertex in the network. We may take $y_v = 0$ as usual, since the problem is balanced. If (w, v) is not an arc of the tree, the vector $\mathbf{y}$ is $\mathbf{0}$. If (w, v) is an arc in the tree, then $y_w = 1$. In this case $\mathbf{y}$ is a vector of 0s and 1s.

Suppose $\{p, q\}$ is an edge in the bipartite graph. If the arc (p, q) is an arc in the tree of the FTS, then $y_p = y_q$. If the FTS is optimal and if (p, q) is not an arc of the tree, then it is not a profitable arc. Hence $y_p \geqslant y_q$. Consequently, $y_p = 1$ or $y_q = 0$.

Thus there are three possibilities for any edge $\{p, q\}$ in the bipartite graph where p is a left vertex and q is a right vertex: (i) $y_p = y_q = 1$; (ii) $y_p = 1$, $y_q = 0$; and (iii) $y_p = 0$, $y_q = 0$. Let W_1 be the set of left vertices p such that $y_p = 1$, and let W_2 be the set of right vertices q such that $y_q = 0$. Then the union W of these two sets is indeed a covering. Thus, using the dual solution of the optimal solution of the transportation problem, we can obtain a covering in the bipartite graph.

We now have the setting to prove Konig's theorem.

Theorem 2.8 (Konig's theorem)
In any bipartite graph, the size of a maximum matching equals the size of a minimum covering.

Proof

If the bipartite graph has no edges, then the theorem is obviously true. Assume it has at least one edge, which implies that there is a matching of cardinality of at least 1. Let M be an optimal matching obtained by solving the transshipment problem using the network simplex method, and let W be the covering defined by this optimal solution. Since the cardinality of M is at least 1, the optimal solution of the transshipment problem will have a positive component along the arc (w, v). Hence the arc (w, v) is an arc of the tree in the optimal FTS which implies $y_w = 1$.

If p is a left vertex in the covering W, by definition $y_p = 1$, which implies that the arc (p, v) is not in the tree. So (since the tree is weakly connected) there exists a right vertex q such that (p, q) is in the tree, implying $y_q = 1$, and consequently q is not in a vertex in W.

Likewise if q is a right vertex in W, $y_q = 0$, implying that the arc (w, q) is not in the tree. So there should be a left vertex p such that (p, q) is in the tree, implying $y_p = 0$, and consequently p is not in the set W.

In other words, every vertex in the covering W is an endpoint of some arc in the optimal tree whose other endpoint is not in the covering. Thus $|M| \geqslant |W|$, which implies $|M| = |W|$.

Finally, if C is any other covering, then $|C| \geqslant |M|$, which implies W is a minimum covering.

Example 2.9

Verify Konig's theorem for the bipartite graph shown below.

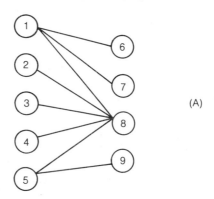

(A)

By inspection of this graph, it is obvious that the set $W = \{1, 5, 8\}$ is a minimum covering consisting of three vertices and the set $M = \{\{1, 6\}, \{4, 8\}, \{5, 9\}\}$ is a maximum matching consisting of three edges.

The tree in an optimal FTS is as shown below. The flow along (w, v) is 3, which is the size of a largest matching. The flows along $(2, v)$, $(3, v)$, $(1, 6)$, $(4, 8)$, $(5, 9)$ and $(w, 7)$ are all equal to 1. The flows along the other arcs are all zero. The components of the dual solution are as follows: $y_v = y_2 = y_3 = y_4 = y_8 = 0$. All other components are 1. The left vertices with y-components equal to 1 are 1 and 5. Vertex 8 is the only one right vertex whose y-component is 0. Thus $W = \{1, 5, 8\}$ is a minimum covering. The arcs in the tree from a left vertex to a right vertex with positive flow constitute a maximum cardinality matching. The arc edges corresponding to these arcs are $\{1, 6\}$, $\{4, 8\}$ and $\{5, 9\}$.

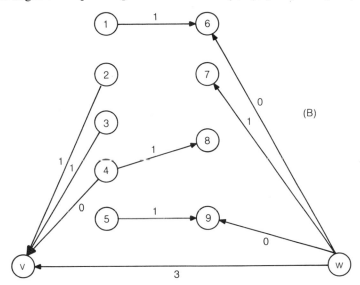

We now establish an equivalent formulation of Konig's theorem in the context of matrices. A 'line' in a matrix is a common term to denote either a row or a column. Suppose P is a property that an element of a matrix may or may not have. A set of elements in the matrix having the property P is said to be **P-independent** if no two of them from the set lie on the same line. The size of the largest P-independent set is called the **P-rank** of the matrix.

Theorem 2.9 (Konig–Egervary theorem)
The P-rank of a matrix is equal to the minimum number of lines that contain all the elements of the matrix with the property P.

Proof
If the matrix is $m \times n$, construct a bipartite graph with m left vertices and n right vertices. Join a left vertex i and a right vertex j by an edge if and only if the (i, j)th element in the matrix has the property P. The desired result follows as an immediate consequence of Theorem 2.8.

Both Konig's theorem and the Konig–Egervary theorem assert the same property. The only difference is that the former is in terms of bipartite graphs, whereas the latter is in terms of matrices.

In the case of a binary matrix, we may say that an element has the property P if it is positive. In this case the P-rank is called the **term-rank** of the matrix.

Example 2.10

Suppose **A** is a binary matrix and an element in **A** has property P if it is positive. State the Konig–Egervary theorem in this context and verify it for the following 5×4 matrix **A**:

$$\mathbf{A} = \begin{bmatrix} 1 & 1 & 1 & 0 \\ 0 & 0 & 1 & 0 \\ 0 & 0 & 1 & 0 \\ 0 & 0 & 1 & 0 \\ 0 & 0 & 1 & 1 \end{bmatrix}$$

The Konig–Egervary theorem for binary matrices is as follows. The maximum number of 1s that can be chosen from a binary matrix such that no two selected 1s lie on the same line is equal to the minimum number of lines needed to cover all the 1s in the matrix.

In the given matrix **A**, we can choose the elements (1, 1), (2, 3) and (5, 4) to constitute a set of three positive elements, no two of which lie in the same line. All the 1s in the matrix can be covered by three lines: one along column 3, one along row 1 and one along row 5. Thus the theorem is verified if we decide that an element has property P if and only if it is positive.

(If each row of the matrix given above corresponds to a left vertex and if each column corresponds to a right vertex, then the matrix **A** of this example is (part of) the adjacency matrix of the bipartite graph of Example 2.9. The (i, j)th element in the matrix is 1 if and only if there is an edge joining the vertex i and the vertex $(5 + j)$. A set of positive elements of maximum cardinality such that no two positive numbers are in the same line is the set $\{(1, 1), (2, 3), (5, 4)\}$ which defines the set $\{\{1, 6\}, \{2, 8\}, \{3, 9\}\}$ of edges, giving a maximum cardinality matching. A set of lines of minimum cardinality which covers all the 1s in the matrix is the set consisting of row 1, row 5 and column 3 and this set corresponds to the set $\{1, 5, 8\}$ of vertices, which is indeed a minimum covering of the graph.)

System of distinct representatives

If it is possible to choose exactly one element from each set belonging to a family of subsets of a finite set X and if the elements thus chosen constitute a set S, then the family has a transversal or a **system of distinct representatives** (SDR) represented by the set S. We say that S is an SDR of the family of sets. If a family has an SDR, then it obviously satisfies the following condition: the number of elements in the union of sets belonging to any subfamily consisting of k sets is at least k. The fact that this condition is sufficient for the existence of an SDR is a consequence of Konig's theorem.

Example 2.11

Consider the family $\{A_1, A_2, A_3, A_4\}$ of four sets where $A_1 = \{a, b, c\}$, $A_2 = \{a, b\}$, $A_3 = \{b, c\}$ and $A_4 = \{a, c\}$. Obviously the family has no SDR. It is easy to see that the sufficiency condition is violated when $k = 4$ because the union of these four sets has only three elements. For an SDR to exist, the condition should hold for $k = 1, 2, 3$ and 4.

Theorem 2.10 (Hall's marriage theorem)
The family $F = \{X_1, X_2, \ldots, X_m\}$ has an SDR if and only if the cardinality of the union of any k sets chosen from the family is at least k for every k ($1 \leqslant k \leqslant m$).

Proof
The condition is obviously necessary.

To prove sufficiency, suppose that the family F has no SDR. Let $X = \{a_1, a_2, \ldots, a_n\}$ be the union of the m sets in F. Construct the bipartite graph (F, X, E) in which there is an edge between the left vertex X_i and the right vertex a_j if and only if a_j is an element of X_i.

Let r be the size of the largest size matching in G, which is also equal to the size of the smallest size covering W by Konig's theorem.

The family F has an SDR if and only if there is a matching of size m in G. So $r < m$ since F has no SDR. Hence there is at least one left vertex which is not in W. Let $X_1, X_2, \ldots, X_k$ be the left vertices which are not in W and let S be the union of these k sets. By hypothesis, the cardinality of S is at least k. Now every element in S is a right vertex which is necessarily an element in W. So the number of right vertices in W is at least k. But the number of right vertices in W is exactly $r - (m - k)$. So $r - (m - k) \geqslant k$ which implies $r \geqslant m$. But $r < m$. This contradiction establishes that the family has an SDR.

(There are several proofs for this theorem – the first by Philip Hall in 1935 – answering the following question known as the **marriage problem**. Suppose there are m men and n women in a party where $n \geqslant m$. Assume that each man is acquainted with at least one woman in the party. Under what conditions, can these men and women be paired as m couples such that in each pair the man is acquainted with the woman? According to the theorem just proved, a necessary and sufficient condition (known as the **marriage condition**) for the solution of the marriage problem is this: every set of q men collectively is acquainted with at least q women for every q.)

In a bipartite graph $G = (X, Y, E)$, a **complete matching** from X to Y is a matching M such that every vertex in X is incident to an edge in M. If both X and Y have the same number of elements, a complete matching from X to Y is a perfect matching.

Theorem 2.11
Let $G = (X, Y, E)$ be a bipartite graph, and, for every subset A of X, let $f(A)$ be the set of those vertices in Y which are joined to at least one vertex in A. Then a complete matching from X to Y exists if and only if $|f(A)| \geqslant |A|$ for every subset A of X.

Proof
This theorem is a graph-theoretic formulation of Hall's marriage theorem and as such is equivalent to Theorem 2.10.

Theorem 2.12
Hall's marriage theorem implies Konig's marriage theorem.

Proof
Suppose $G = (X, Y, E)$ is a regular bipartite graph of degree $r > 0$. Let A be any subset of X and $f(A)$ be the set of vertices in Y which are joined to at least one vertex in A by an edge of the graph. Let E_1 be the set of edges adjacent to the vertices in A, and let E_2 be the set of edges adjacent to the vertices in $f(A)$. Then $|E_1| \leqslant |E_2|$. But $|E_1| = r|A|$ and $|E_2| = r|f(A)|$. Thus the marriage condition $|f(A)| \geqslant |A|$ holds for every subset A of X. So there is a complete matching from X to Y. But both X and Y have the same number of elements since $r > 0$. So there is a perfect matching in the bipartite graph.

Theorem 2.13
Hall's marriage theorem implies the Konig–Egervary theorem.

Proof
Let $\mathbf{B}$ be any $n \times m$ matrix. If an element in $\mathbf{B}$ has property P, replace it by 1. Otherwise replace it by 0. So, without loss of generality, we can take $\mathbf{B}$ as a binary matrix. A set of lines which cover all the 1s in $\mathbf{B}$ can be considered as a covering in $\mathbf{B}$. We have to show that the term-rank of $\mathbf{B}$ is equal to the smallest size of covering in $\mathbf{B}$.

Notice that if two 1s are to be accounted for while computing the term-rank, these two 1s cannot lie in the same line. So at least two lines are needed to cover these two 1s. In other words, the term-rank cannot exceed the size of any covering. Thus $p \leqslant q$, where p is the term-rank and q is the cardinality of the smallest covering.

Let the q lines in a smallest covering consist of r rows and s columns $(r + s = q)$ and, without loss of generality, assume that these are the first r rows and s columns of the matrix $\mathbf{B}$. Then the $(n - r) \times (m - s)$ submatrix in the lower right-hand corner is the zero matrix.

Let $A_i = \{ j : j > s \text{ and } b_{ij} = 1 \}$ where $i = 1, 2, \ldots, r$. If A_i is empty then row i can be pushed to the bottom of $\mathbf{B}$. But this cannot be done because of the minimality of the covering. So no set A_i is empty.

Suppose the cardinality of the union of any k sets from the family $\{ A_i \}$ is k'. If $k' < k$, then it is easy to see that we can obtain a new covering of the matrix by replacing the r rows by $r - k$ rows and the s columns by $s + k'$ columns. The cardinality of the new covering is

$$(r - k) + (s + k') = r + s + (k' - k) = q + (k' - k) < q$$

This inequality violates the minimality. Thus the union of any k sets from the family $\{ A_i \}$ will have at least k elements. Thus, by Hall's marriage theorem, it is possible to choose r 1s from the $r \times (m - s)$ submatrix in the upper right-hand corner of $\mathbf{B}$ such that no two 1s lie on the same line.

Similarly, we can choose s 1s from the $(n - r) \times s$ submatrix in the lower left-hand corner of **B** such that no two 1s lie on the same line.

Thus $r + s$ cannot exceed the term-rank p. In other words, $q \leqslant p$ which implies $p = q$.

Partially ordered sets

A **partially ordered set** or **poset** $P = (X, \leqslant)$ is a set X on which an order relation $\leqslant$ is defined satisfying the following three properties:

(i) Reflexivity: $x \leqslant x$ for every x in X.
(ii) Symmetry: $x \leqslant y$ and $y \leqslant x$ imply $x = y$.
(iii) Transitivity: $x \leqslant y$ and $y \leqslant z$ imply $x \leqslant z$.

Some examples of partially ordered sets are the following: the family of subsets of a set, where $\leqslant$ means set inclusion; the set of divisors of a positive integer, where $x \leqslant y$ means x divides y; and the field of complex numbers, where $a + ib \leqslant c + id$ means $a \leqslant c$ and $b \leqslant d$.

Two elements x and y in a poset are **comparable** if $x \leqslant y$ or $y \leqslant x$. They are **incomparable** otherwise. A subset C of a poset P is called a **chain** (or a **linear order**) if every pair in C is a comparable pair. A set of n elements in a poset is a chain if and only if these elements can be enumerated as $y_i (i = 1, 2, \ldots, n)$ such that $y_1 \leqslant y_2 \leqslant y_3 \leqslant \cdots \leqslant y_n$. A subset of P, on the other hand, is an **antichain** if every pair in it is an incomparable pair.

Suppose a poset P is partitioned into q subsets such that each such subset is a chain. In other words, we have a **chain decomposition** of the poset. An antichain D in the poset can contain at most one element from each of these chains. So the cardinality of D cannot exceed q. Thus the cardinality of a largest antichain is less than or equal to the minimum number of disjoint chains into which the poset can be partitioned. Dilworth's theorem asserts that these two numbers are equal. An elegant proof of this theorem (given below) using the Konig–Egervary theorem is due to Dantzig and Hoffman (1956).

Theorem 2.14 (**Dilworth's theorem**)
In a finite poset the maximum size of an antichain is equal to the minimum number of (disjoint) chains into which the poset can be partitioned.

Proof
The finite poset $P = (\{x_1, x_2, \ldots, x_n\}, \leqslant)$ can be represented by an $n \times n$ binary matrix $\mathbf{A} = [a_{ij}]$, where $a_{ij} = 1$ if and only if $x_i < x_j$. Hence if $a_{ij} = 1$, then $a_{ji} = 0$.

For the sake of simplicity, a set of 1s in the matrix $\mathbf{A}$ is called an 'independent set' if no two 1s in the set are on the same line. Every chain in P consisting of two or more elements defines an independent set.

Suppose a chain decomposition of P consists of p_i nonsingleton chains of length $k_i (i = 1, 2, \ldots, r)$ and q singleton chains of length l. Then $n = \Sigma p_i q_i + q$. Now this decomposition will define an independent set in which the number of

1s will be m, where

$$m = p_1(q_1 - 1) + p_2(q_2 - 1) + \cdots + p_r(q_r - 1) = \Sigma p_i q_i - \Sigma p_i$$

Thus $m = n - (p + q)$, where $p = \Sigma p_i$.

Since $p + q$ is the total number of chains in the decomposition, we can now say that the number of chains in a decomposition plus the number of 1s in the corresponding independent set equals the cardinality of the poset. It follows that if there exists a maximum independent set of size t (i.e. if the term-rank of **A** is t), then there is a minimum chain decomposition of q singleton chains and $n - t - q$ chains of length greater than 1. Then, by the Konig–Egervary theorem, the matrix can be covered by t lines, and cannot be covered by less than t lines. This minimal cover corresponds to a set D of $n - t - q$ elements of P (one from each of the $n - t - q$ nonsingleton chains).

The set D' of elements which constitute the singleton chains is of cardinality q. Then the union of D and D' is a set of noncomparable elements of cardinality $n - t$. So there exist an antichain of cardinality $n - t$ and a chain decomposition of the poset consisting of $n - t$ chains. This completes the proof.

Each poset P can be represented by a directed graph G in which each vertex corresponds to an element of the poset and an arc is drawn from x to y if $x \leqslant y$. If we delete arcs that are present due to transitivity, we get a subgraph H of the digraph known as the **Hasse diagram** of the poset: if $x \leqslant y$ and $y \leqslant z$, then the arc from x to z does not appear in the Hasse diagram.

Example 2.12
Verify Dilworth's theorem for the poset which is represented by the Hasse diagram given below.

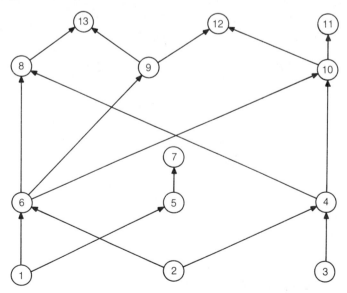

The chains $\{1, 5, 7\}$, $\{2, 6, 8, 13\}$, $\{3, 4, 10, 11\}$ and $\{9, 12\}$ constitute a chain decomposition of the poset into four chains. The set $\{7, 8, 9, 10\}$ is an antichain of cardinality 4.

If a poset has a disjoint partition consisting of p chains, then no antichain can have cardinality greater than p. A partition with p chains is a **saturated partition** if there exists an antichain of cardinality p. Dilworth's theorem asserts that every poset has at least one saturated partition.

Suppose G is an acyclic digraph, i.e., a digraph with no directed cycles. The set E of arcs in a digraph G can be considered as a partially ordered set $(E, \leqslant)$ if $e \leqslant f$ means that there exists a directed path (in the digraph) which contains the arc e and the arc f. We then have the following graph-theoretic version of Dilworth's theorem: the maximum number of arcs in a set A of arcs of an acyclic digraph, no two of which are in the same directed path, is equal to the minimum number of directed paths whose union contains the set A.

Theorem 2.15
Dilworth's theorem implies Hall's marriage theorem.

Proof
Let $\{A_i : i = 1, 2, \ldots, n\}$ be a family of subsets of the set $E = \{x_1, x_2, \ldots, x_m\}$ satisfying Hall's marriage condition. We have to show that the family has an SDR.

On the set $X = \{x_1, x_2, \ldots, x_m, A_1, A_2, \ldots, A_n\}$, let $<$ be a strict partial order, where $x_i < A_j$ if x_i is an element of A_j. In the poset X, the set E is an antichain of cardinality m.

Now consider an arbitrary antichain D in X consisting of p elements from E and q sets from the family. We may write $D = \{x_1, x_2, \ldots, x_p, A_1, A_2 \ldots, A_q\}$. None of the p elements x_i from D can be an element of the union of these q sets in D. So the union can have at most $m - p$ elements. But this union of q sets has at least q elements by the marriage condition. Thus $q \leqslant m - p$, implying that $p + q \leqslant m$, which shows that E is a maximal antichain in X.

So by Dilworth's theorem there exists a partition of X into m chains. Of these, n chains necessarily consist of two elements, one from E and one from the family. By a suitable reindexing of X, we can notate the chains as $x_i < A_i$. In other words, the set $\{x_1, x_2, \ldots, x_n\}$ is an SDR for the family.

Theorem 2.16 (Mirsky's dual version of Dilworth's theorem)
In a finite poset the maximum size of a chain is equal to the minimum number of (disjoint) antichains into which the poset can be partitioned.

Proof
Suppose the number of elements in a chain is p and suppose the poset is partitioned into q antichains. Then $p \leqslant q$. So it is enough if we show that if m is

the number of elements in a largest chain, then there exists a partition of the poset into m antichains. We prove this by induction on m. The theorem is true when $m = 1$. Suppose it holds for $m - 1$.

Let P be a poset which has a largest chain consisting of m elements. An element x is maximal if $x \leqslant y$ implies $x = y$. The set X of all maximal elements in P is an antichain. In the subposet $P - X$, the number of elements in a largest chain is at most $n - 1$. The length of a largest chain in $P - X$ is $m - 1$. So by the induction hypothesis, $P - X$ can be partitioned into $m - 1$ pairwise disjoint antichains, which implies that P can be partitioned into m disjoint chains. So the theorem is true for m as well.

Example 2.13
Verify Mirsky's theorem for the poset whose Hasse diagram is as in Example 2.12.

No chain in the poset contains more than four elements. A chain of four elements is $\{1, 6, 8, 13\}$. The four antichains $\{1, 2, 3\}$, $\{4, 5, 6\}$, $\{7, 8, 9, 10\}$ and $\{11, 12, 13\}$ constitute a partition of the poset.

Summary: implications and equivalences

1. The integrality theorem (in section 2.2) asserts that if the supply–demand vector in a feasible transshipment is an integer vector, then the problem has an optimal solution which is also an integer vector.
2. The integrality theorem implies Konig's marriage theorem (Theorem 2.5), the Birkhoff–von Neumann theorem (Theorem 2.7) and Konig's theorem (Theorem 2.8).
3. Konig's theorem and the Konig–Egervary theorem (Theorem 2.9) are equivalent.
4. Konig's theorem implies Hall's marriage theorem (Theorem 2.10).
5. Hall's marriage theorem implies Konig's marriage theorem (Theorem 2.12) and the Konig–Egervary theorem (Theorem 2.13).
6. Konig's theorem implies Dilworth's theorem (Theorem 2.14).
7. Dilworth's theorem implies Hall's marriage theorem (Theorem 2.15).
8. Thus Konig's theorem, the Konig–Egervary theorem, Hall's marriage theorem and Dilworth's theorem are equivalent.

(This discussion will continue in Chapter 4.)

2.6 ASSIGNMENT PROBLEMS

The assignment problem is a special kind of transportation problem in which the supply at each source and the demand at each sink are both equal to one indivisible unit. The integrality theorem again plays a crucial role. Moreover, the assumption that the total supply is equal to the total demand implies that the number of sources equals the number of sinks in an assignment problem. If this is not the case, then we introduce artificial vertices and artificial arcs as before to correct this lack of balance.

The assignment problem can be either a minimization problem or a maximization problem. For example if sink j (with demand 1) denotes the city to which a firm sends a new recruit represented by source i and if c_{ij} is the salary to be given to i if i is assigned to j, then from the point of view of the firm the objective is to assign the new employees to their job sites such that the total wage to be paid is minimized. On the other hand, if the sink j denotes the type of a job in the firm and if c_{ij} is the score obtained by the new recruit i while testing for that job, the objective, in the best interest of the firm, is to assign these new recruits to jobs such that the total of all the scores is a maximum. Another optimization problem of interest is the problem of matching as many sources as possible with as many sinks. This is an assignment (maximization) problem in which $c_{ij} = 1$ for each arc (i, j) in the graph.

It is obvious that the problem 'maximize $\mathbf{c} \cdot \mathbf{x}$' is equivalent to the problem 'minimize $-\mathbf{c} \cdot \mathbf{x}$'. Thus we can always use the network simplex method to solve the assignment problem. In any $n \times n$ assignment problem, every feasible tree solution is degenerate since out of the $2n - 1$ components of a feasible tree solution exactly n are 1s and the remaining components are 0s. Here is a simple example.

Example 2.14
Suppose the cost matrix is

$$\mathbf{c} = \begin{bmatrix} 5 & 4 & 2 & 4 \\ 6 & 3 & 3 & 5 \\ 6 & 2 & 5 & 2 \\ 6 & 3 & 2 & 7 \end{bmatrix}$$

Using the northwest corner method, we obtain a feasible tree solution with positive flow components of one unit each along the arcs $(1, 5)$, $(2, 6)$, $(3, 7)$ and $(4, 8)$, which are the diagonal elements in the tableau. The initial cost, 20, is the sum of the diagonal elements of the cost matrix. Since there are eight vertices, any FTS should have seven components. So we take three more arcs to complete a tree. In an assignment problem, such arcs with zero flow are usually chosen on one side of the diagonal, either above it or below. Let us take the three cells $(1, 6)$, $(2, 7)$ and $(3, 8)$ to represent the other three arcs in the tree.

We begin with the following tableau:

5	4	2	4		
1	0			1	−1
6	3	3	5		
	1	0		1	0
6	2	5	2		
		1	0	1	−2
6	3	2	7		
			1	1	−7
1	1	1	1	4	
4	3	3	0		

The arc (4, 7) is a profitable arc for this FTS. When (4, 7) enters, either (3, 7) or (4, 8) leaves the tree. If we decide that (4, 8) leaves the tree, we get the following updated tableau:

5	4	2	4		
1	0			1	−1
6	3	3	5		
	1	0		1	0
6	2	5	2		
		0	1	1	−2
6	3	2	7		
		1		1	1
1	1	1	1	4	
4	3	3	0		

The arc (1, 7) is profitable and enters the tableau. Either the arc (1, 6) or the arc (2, 7) can leave. Suppose (1, 6) leaves. The tableau is updated as follows:

5	4	2	4		
1		0		1	−1
6	3	3	5		
	1	0		1	0
6	2	5	2		
		0	1	1	−2
6	3	2	7		
		1		1	1
1	1	1	1	4	
4	3	3	0		

The arc (3, 5) enters. Either (3, 7) or (1, 7) can leave. If the arc (3, 7) leaves we have the following tableau:

5	4	2	4		
1		0		1	−1
6	3	3	5		
	1	0		1	−2
6	2	5	2		
0			1	1	−2
6	3	2	7		
		1		1	−1
1	1	1	1	4	
4	1	1	0		

The arc (3, 6) enters. The arc (2, 7) leaves giving the following tableau:

5 (1)	4	2 (0)	4	1	−1
6	3 (1)	3	5	1	−3
6 (0)	2 (0)	5	2 (1)	1	−2
6	3	2 (1)	7	1	−1
1	1	1	1	4	
4	0	1	0		

The arc (2, 5) enters and the arc (3, 5) leaves giving the following tableau:

5 (1)	4	2 (0)	4	1	−2
6 (0)	3 (1)	3	5	1	−3
6	2 (0)	5	2 (1)	1	−2
6	3	2 (1)	7	1	−2
1	1	1	1	4	
3	0	0	0		

In this tableau there are no profitable arcs. So the FTS given in this tableau is optimal. The arcs with positive flow in the optimal assignment are $(1, 5)$, $(2, 6)$, $(3, 8)$ and $(4, 8)$ giving a total weight of $5 + 3 + 2 + 2 = 12$. In other words, we choose the $(1, 1)$th, $(2, 2)$th, $(3, 4)$th and $(4, 3)$th elements of the cost matrix. In this formulation the optimal assignment is represented as $1 \rightarrow 1$, $2 \rightarrow 2$, $3 \rightarrow 4$ and $4 \rightarrow 3$.

The Hungarian method

Due to the integrality theorem and the special structure of the supply–demand vector, we can solve the assignment problem by an easier method known as the **Hungarian method**, which involves modifying the cost matrix of the problem so that it can be 'matched' with a permutation of the identity matrix.

Consider the assignment (minimization) problem and the associated complete bipartite graph with n sources and n sinks. Any feasible flow is an $n \times n$ binary matrix in which the element 1 appears exactly once in each row and in each column. In other words, a feasible flow is a permutation of the $n \times n$ identity matrix and vice versa. Suppose the cost matrix c is nonnegative with at least n zeros. If there exists a permutation P of the identity matrix such that the n 1s of P lie in the same position as n of the 0s of c, then obviously the assignment given by P is an optimal assignment. In this case we say that c is matched with P. If the given matrix c does not have this property, we try to see whether c can be modified towards this end.

Theorem 2.17
Suppose c is the cost matrix and c' is the matrix obtained by adding a number t to each element in the ith row or to each element in the ith column. Then a solution is optimal with respect to c' if and only if it is optimal with respect to c.

Proof
Let $u = [u_{ij}]$ be any feasible solution with respect to c'. Then $c' \cdot u = c \cdot u + t$. So u is optimal with respect to c' if and only if it is optimal with respect to c.

Thus given an arbitrary nonnegative cost matrix, we first see whether by systematically subtracting positive numbers from rows and columns we can modify c to form a new nonnegative matrix c' which matches with a permutation of the identity matrix. If we succeed in this effort, then we obtain an optimal solution without the help of the network simplex method. Notice that by subtracting the smallest number of a row from each element of that row and by doing it for all rows we have a modified matrix in which each row has a 0. Then we carry out the same procedure for each column, ending up with a modified matrix c' which has a 0 in each row and in each column. We are done if c' matches with a permutation of the identity matrix.

Example 2.15
Consider the cost matrix in the assignment (minimization) problem in Example 2.14. In this example, we subtract 2 from row 1, 3 from row 2, 2 from row 3 and 2 from row 4. In the first column of the modified matrix all elements are still positive. So we subtract from each element in the first column the smallest element of that column. Thus we have a modified matrix c' which matches with

a permutation matrix $\mathbf{P}$ as follows:

$$\mathbf{c}' = \begin{bmatrix} 0 & 2 & 0 & 2 \\ 0 & 0 & 0 & 2 \\ 1 & 0 & 3 & 0 \\ 1 & 1 & 0 & 5 \end{bmatrix} \quad \mathbf{P} = \begin{bmatrix} 1 & 0 & 0 & 0 \\ 0 & 1 & 0 & 0 \\ 0 & 0 & 0 & 1 \\ 0 & 0 & 1 & 0 \end{bmatrix}$$

If $S = \{1, 2, 3, 4\}$ is the set of sources and $D = \{5, 6, 7, 8\}$ is the set of sinks, the nonzero components of the optimal solution are along the arcs $(1, 5)$, $(2, 6)$, $(3, 8)$ and $(4, 7)$. In the matrix setting, the optimal assignment is $1 \to 1$, $2 \to 2$, $3 \to 4$ and $4 \to 3$, with a total cost of $c_{11} + c_{22} + c_{34} + c_{43} = 5 + 3 + 2 + 2 = 12$.

We cannot rush to the conclusion that this method of solving the problem will immediately lead us to an optimal solution, for it is not at all necessary that after performing these subtractions from the elements in the matrix, the modified $n \times n$ matrix will match with a permutation of the identity matrix.

For example, in the following 4×4 matrix:

$$\mathbf{c}' = \begin{bmatrix} 2 & 0 & 3 & 4 \\ 4 & 3 & 0 & 2 \\ 0 & 0 & 3 & 0 \\ 2 & 3 & 0 & 1 \end{bmatrix}$$

the only zero of row 2 and the only zero of row 4 are both in column 3, showing that there is no permutation with which $\mathbf{c}'$ can be matched. In order to obtain a matching with a permutation of the identity matrix, we have to modify the cost matrix again such that there are enough 0s in the right place.

To summarize, there are three major steps involved in the Hungarian method:

Step 1. Perform the subtractions in the cost matrix $\mathbf{c}$ such that we have a modified cost matrix $\mathbf{c}'$.

Step 2. Check whether $\mathbf{c}'$ can be matched with a permutation of the identity matrix. If the answer is yes, an optimal assignment is easily discernible by this matching and we stop. Otherwise we go to step 3.

Step 3. Redistribute the zeros in $\mathbf{c}'$ and go back to step 2.

We now define an efficient way of carrying out step 2.

Step 2a. Locate a row or a column in the modified matrix $\mathbf{c}'$ with exactly one 0. If there is no such row or column, locate a row or column with the smallest number of 0s. Tie-breaking is arbitrary. If a row is located, we draw a vertical line through the 0. If a column is located, we draw a horizontal line through the 0.

Step 2b. Repeat the previous step (while not taking into consideration any 0 with a line through it) till every 0 in the matrix has at least one line through it.

Step 2c. If the number of lines drawn is n, the matrix can be matched with a permutation of the $n \times n$ identity matrix. The 0s which are identified for the drawing of these lines will point out an optimal solution. If the number of lines is less than n, we go to step (3).

Here are two 5×5 matrices, **A** and **B**. The procedure explained in step 2 gives five lines in **A** giving an optimal solution right away. But in **B**, this method does not give five lines.

$$\mathbf{A} = \begin{bmatrix} 4 & 1 & 3 & 2 & 4 \\ 6 & 2 & 2 & 4 & 5 \\ 1 & 3 & 4 & 1 & 1 \\ 5 & 2 & 3 & 4 & 1 \\ 7 & 6 & 5 & 3 & 3 \end{bmatrix} \quad \mathbf{B} = \begin{bmatrix} 4 & 9 & 3 & 11 & 4 \\ 9 & 8 & 3 & 10 & 8 \\ 7 & 5 & 3 & 8 & 6 \\ 9 & 5 & 3 & 4 & 6 \\ 10 & 11 & 7 & 10 & 11 \end{bmatrix}$$

The modified matrices obtained from **A** and **B** are:

$$\mathbf{A}' = \begin{bmatrix} 3 & 0 & 2 & 1 & 3 \\ 4 & 0 & 0 & 2 & 3 \\ 0 & 2 & 3 & 0 & 0 \\ 3 & 1 & 2 & 3 & 0 \\ 4 & 3 & 2 & 0 & 0 \end{bmatrix} \quad \mathbf{B}' = \begin{bmatrix} 0 & 4 & 0 & 7 & 0 \\ 5 & 3 & 0 & 6 & 4 \\ 3 & 0 & 0 & 4 & 2 \\ 5 & 0 & 0 & 0 & 2 \\ 2 & 2 & 0 & 2 & 3 \end{bmatrix}$$

In **A**', we first locate row 1 which has a unique zero as the $(1, 2)$th entry. So we draw a vertical line (line 1) through column 2. When this line is drawn, row 2 has a unique zero as the $(2, 3)$th entry. A vertical line (line 2) is drawn through column 3. Then row 4 is located which has a unique zero as the $(4, 5)$th entry. So a vertical line (line 3) is drawn through column 5. At this stage row 5 has a unique zero as the $(5, 4)$th entry. So a vertical line (line 4) is drawn through column 4. Finally, there is a unique zero as the $(3, 1)$th element in column 1. So a horizontal line (line 5) is drawn through row 3. The identified entries are the $(1, 2)$th, $(2, 3)$th, $(4, 5)$th, $(5, 4)$th and $(3, 1)$th entries in the matrix. An optimal assignment is $1 \to 2, 2 \to 3, 3 \to 1, 4 \to 5$ and $5 \to 4$ with a total weight of $1 + 2 + 1 + 1 + 3 = 8$.

But to cover the zeros in the 5×5 matrix **B**', we need only four lines through row 1, row 4, column 2 and column 3. So we are unable to obtain an optimal assignment for **B** using the modified matrix.

How do we solve the assignment problem if the number of lines needed to cover the 0s of the modified $n \times n$ cost matrix is less than n? This takes us to the last step in the Hungarian method. As we saw in section 2.5 (Theorem 2.9), the maximum number of assignments that can be made using the 0s of the matrix is equal to the minimum number of lines that are required to cover the 0s of the matrix. So in a situation like this where we need 0s in the right place, we have to modify the cost matrix further to obtain an optimal assignment.

Suppose the number of lines needed to cover the zeros is k which is less than n. Let m be the smallest uncovered element. The cost matrix is further modified using m as follows. We subtract m from the elements of each of the uncovered rows. A zero in an uncovered row (which is necessarily an element in some covered column) now takes the value $-m$. Now add m to the elements of each of the covered columns. At this stage we remove all the lines, resulting in a matrix which can be matched with a permuation of the identity matrix.

The modification procedure given above can be stated in a simpler form as follows. A typical element in the modified cost matrix at the end of step (2) belongs to one of the three categories: (a) it is uncovered; (b) it is doubly covered in the sense that it is at the point of intersection of a vertical line and a horizontal line; and (c) it is covered but not doubly covered. Now we subtract m from each uncovered element and add m to each doubly covered element. The other elements are left alone. The resulting matrix is the modified matrix we are looking for.

In the matrix $\mathbf{B}'$ above, the smallest uncovered element is 2, which is subtracted from each uncovered element. There are nine uncovered elements in the matrix. The doubly covered elements are the (1, 2)th, (1, 3)th, (3, 3)th and (5, 3)th entries. We add 2 to these four elements. We then obtain the following matrix:

$$\mathbf{B}'' = \begin{bmatrix} 0 & 6 & 2 & 7 & 0 \\ 3 & 3 & 0 & 4 & 2 \\ 1 & 0 & 0 & 2 & 0 \\ 5 & 2 & 2 & 0 & 2 \\ 0 & 2 & 0 & 0 & 1 \end{bmatrix}$$

This matrix can be matched with the following permutation $\mathbf{P}$ of the 5×5 identity matrix:

$$\mathbf{P} = \begin{bmatrix} 0 & 0 & 0 & 0 & 1 \\ 0 & 0 & 1 & 0 & 0 \\ 0 & 1 & 0 & 0 & 0 \\ 0 & 0 & 0 & 1 & 0 \\ 1 & 0 & 0 & 0 & 0 \end{bmatrix}$$

Thus an optimal assignment for the cost matrix $\mathbf{B}$ is $1 \rightarrow 5$, $2 \rightarrow 3$, $3 \rightarrow 2$, $4 \rightarrow 4$ and $5 \rightarrow 1$, with cost $4 + 3 + 5 + 4 + 10 = 26$.

The assignment problem is also known as the bipartite matching problem. We will discuss the assignment problem in a more general setting in Chapter 5, when we study optimal matchings in bipartite and nonbipartite graphs.

2.7 EXERCISES

1. If $\mathbf{y}'$ is a dual solution of the feasible tree solution $\mathbf{x}'$ of the transshipment problem with cost vector $\mathbf{c}$, constraint matrix $\mathbf{A}$ and supply–demand vector $\mathbf{b}$, prove $\mathbf{c}\mathbf{x}' = \mathbf{y}'\mathbf{b}$. (Here $\mathbf{A}$ is $n \times m$, $\mathbf{c}$ is $1 \times m$ and $\mathbf{b}$ is $n \times 1$.)

2. Consider the transshipment problem with the following input:

$$c = [8 \ 1 \ 7 \ 1 \ 1 \ 1 \ 2 \ 2]$$
$$b^T = [-6 \ -9 \ 0 \ 0 \ 7 \ 8]$$

$$A = \begin{bmatrix} -1 & -1 & -1 & 0 & 0 & 0 & 0 & 0 \\ 1 & 0 & 0 & -1 & 0 & 0 & 0 & 0 \\ 0 & 0 & 0 & 0 & -1 & 1 & 0 & 0 \\ 0 & 1 & 0 & 1 & 0 & -1 & -1 & -1 \\ 0 & 0 & 1 & 0 & 0 & 0 & 1 & 0 \\ 0 & 0 & 0 & 0 & 1 & 0 & 0 & 1 \end{bmatrix}$$

(i) Find a dual solution y corresponding to the feasible tree solution $x^T = [0 \ 6 \ 0 \ 9 \ 8 \ 8 \ 7 \ 0]$.
(ii) Verify that $cx = b^T y$.
(iii) Test whether x is an optimal solution.

3. Suppose $G = (V, A)$ is the acyclic weakly connected digraph with V consisting of vertices v_i ($i = 1, 2, \ldots, 8$) in which the seven arcs are $(v_1, v_2), (v_3, v_2),$ $(v_4, v_3), (v_7, v_2), (v_3, v_6), (v_5, v_6)$ and (v_8, v_7). Relabel the vertices and arcs such that when the last row of the incidence matrix is deleted, the truncated matrix is upper triangular and nonsingular.

4. Solve the transshipment problem with the following input using the network simplex method:
   (i) $V = \{1, 2, \ldots, 8\}$ is the set of vertices.
   (ii) The arcs $e_i (i = 1, 2, \ldots, 12)$ in the digraph are taken sequentially as $(1, 6),$ $(1, 7) (2, 1), (2, 3), (3, 8), (4, 6), (5, 4), (6, 5), (7, 3), (7, 5), (8, 2)$ and $(8, 4)$.
   (iii) $b^T = [-12 \ -4 \ -2 \ 0 \ 6 \ 6 \ 2 \ 4]$, where b is the supply–demand vector.
   (iv) $c = [12 \ 6 \ 3 \ 9 \ 0 \ 3 \ 3 \ 0 \ 6 \ 9 \ 0 \ 3]$ is the cost vector.
   Verify that $cx = yb$ at each stage of the iteration.

5. Suppose $c = [4 \ -2 \ 1 \ 2 \ -3 \ 1 \ 2 \ 4 \ 2 \ 3 \ 1 \ 1]$ in Exercise 4 above. Show that the problem is unbounded. (Since the cost vector has negative components, the possibility that there could be a feasible solution with an arbitrary small cost cannot be ruled out. To show that this is indeed the case it is enough if we produce a profitable arc e such that every arc in the unique cycle created by e is a forward arc, in which case t can be taken as an arbitrary large positive number giving a feasible flow (not an FTS) for which the cost is arbitrarily small.)

6. Suppose $b^T = [-\ 3 \ -9 \ 3 \ 0 \ 3 \ -9 \ 6 \ 9]$ in Exercise 4 above. By enlarging the network, show that the problem is infeasible.

7. Consider the transshipment problem in a network with seven vertices labelled as $1, 2, \ldots, 7$, with $b^T = [-6 \ -4 \ -8 \ 0 \ 0 \ 11 \ 7]$, and nine arcs $(1, 2), (1, 4), (2, 3), (2, 5), (3, 5), (4, 6), (5, 4), (5, 7)$ and $(6, 7)$. Obtain an FTS by enlarging the network with the construction of two additional arcs $(1, 5)$ and $(5, 6)$ so that the entire supply is sent to vertex 5 from each source and

the entire demand is sent from vertex 5 to each sink. After obtaining an FTS, solve the problem with $c = [5\ 9\ 3\ 8\ 2\ 7\ 6\ 3\ 8]$.

8. By constructing appropriate artificial arcs, show that the problem with $x^T = [x_{12}\ x_{14}\ x_{23}\ x_{25}\ x_{35}\ x_{45}\ x_{46}\ x_{57}\ x_{67}]$ and $b^T = [-6\ -4\ -8\ 0\ 0\ 11\ 7]$ is infeasible.

9. By constructing artificial arcs (2, 5) and (5, 8), obtain an FTS where $b^T = [-6\ -8\ -10\ 0\ 0\ 0\ 10\ 8\ 6]$ in the network with arcs (1, 2), (1, 4), (1, 5), (2, 3), (3, 5), (3, 6), (4, 7), (5, 4), (5, 7), (5, 9), (6, 5), (6, 9), (8, 7) and (8, 9).

10. Let $x^T = [x_{13}\ x_{21}\ x_{32}\ x_{51}\ x_{52}\ x_{61}\ x_{64}\ x_{67}\ x_{74}\ x_{75}]$, $c = [1\ 1\ 1\ 1\ 5\ 11\ 20\ 5\ 5\ 6]$ and $b^T = [0\ 2\ 3\ 5\ -2\ -3\ -5]$. By taking $x^T = [3\ 3\ 0\ 0\ 5\ 0\ 3\ 0\ 2\ 3]$ as an initial FTS, solve this problem by network simplex method.

11. Consider the network G with eight vertices $1, 2, \ldots, 8$ and with 12 arcs (1, 2), (2, 6), (3, 2), (3, 4), (4, 1), (4, 8), (5, 1), (6, 5), (7, 3), (7, 6), (8, 5), (8, 7). Suppose the supply–demand vector is $b^T = [-3\ 2\ 3\ 7\ 4\ 5\ 4\ 6]$ and the cost vector is $c = [2\ 3\ 2\ 4\ 2\ 5\ 2\ 6\ 2\ 5\ 2\ 3]$. It can be seen by inspection that the problem can be decomposed. Solve the transshipment problem by decomposing it.

12. Solve the integer linear programming problem:
   Minimize $z = cx$, $Ax = b$, $x \geqslant 0$, where

   (i) $c = [8\ 9\ 5\ 6\ 7]$
   $$b^T = [-4\ -9\ 7\ 6]$$
   $$A = \begin{bmatrix} -1 & 1 & 0 & 0 & 0 \\ 0 & -1 & -1 & -1 & 0 \\ 1 & 0 & 1 & 0 & -1 \\ 0 & 0 & 0 & 1 & 1 \end{bmatrix}$$

   (ii) $c = [2\ 3\ 4\ 4\ 3\ 2]$
   $$b^T = [-4\ -6\ 0\ 3\ 7]$$
   $$A = \begin{bmatrix} -1 & -1 & 0 & 0 & 0 & 0 \\ 0 & 0 & -1 & -1 & 0 & 0 \\ 1 & 0 & 1 & 0 & -1 & -1 \\ 0 & 1 & 0 & 0 & 1 & 0 \\ 0 & 0 & 0 & 1 & 0 & 1 \end{bmatrix}$$

13. Solve the integer linear programming problem with the cost vector $[1\ 1\ 1\ 1\ 1]$ and the constraints as in Example 2.6.

14. Solve the following ILP problem by converting it into a transshipment problem:

$$\text{Minimize } z = 2x_1 + 8x_2 + 3x_3 + 6x_4 + 7x_5$$
$$\text{where } -x_1 + x_2 \qquad\qquad\qquad \geqslant -4$$
$$-x_2 - x_3 - x_4 \qquad \geqslant -9$$
$$x_1 \qquad + x_3 \qquad\quad -x_5 = 7$$
$$x_4 + x_5 = 2$$

(All the variables are nonnegative.)

15. The six constraints of a certain ILP involving 11 nonnegative variables are given below. Obtain the constraint matrix $\mathbf{A}$ and the supply–demand vector $\mathbf{b}$ of a transshipment problem to which the ILP can be converted such that we have a linear system $\mathbf{Ax} = \mathbf{b}$, where $\mathbf{A}$ is the incidence matrix of a digraph.

$$x_1 + x_2 + x_3 - x_4 - x_9 = 0$$
$$x_4 + x_5 + x_6 + x_7 - x_{10} \geqslant 0$$
$$x_1 + x_5 + x_{11} = 6$$
$$x_2 + x_6 + x_8 = 10$$
$$x_3 + x_7 - x_8 \leqslant 8$$
$$x_9 + x_{10} + x_{11} \leqslant 9$$

16. Solve the ILP with $\mathbf{c} = [3\ 1\ 0\ 0\ 1\ 1\ 4\ 2\ 1]$ and the following seven constraints:

$$x_1 + x_2 - x_4 = 2$$
$$x_1 - x_3 + x_8 = -1$$
$$x_3 - x_4 - x_5 = 2$$
$$x_2 - x_6 - x_7 = -6$$
$$-x_6 + x_8 \leqslant -1$$
$$x_5 - x_9 \geqslant 0$$
$$x_7 + x_9 \geqslant 3$$

where all the variables are nonnegative integers.

17. Every transportation problem is a transshipment problem. Show that every transshipment problem can be converted into a transportation problem.

18. Convert the following transshipment problem into a transportation problem and solve it.

$$V = \{1, 2, 3, 4, 5, 6, 7\}$$
$$\mathbf{b}^T = [0\ 2\ 3\ 5\ -2\ -3\ -5]$$
$$\mathbf{x}^T = [x_{13}\ x_{21}\ x_{32}\ x_{51}\ x_{52}\ x_{61}\ x_{64}\ x_{67}\ x_{74}\ x_{75}]$$
$$\mathbf{c} = [1\ 1\ 1\ 1\ 5\ 11\ 20\ 5\ 5\ 6]$$

19. In the following nonoptimal table, it is known that the arc (S_3, D_4) enters the tree in the current FTS in the iteration process. The cost matrix entries are not displayed. Obtain the arc which will leave the tree.

	D_1	D_2	D_3	D_4	D_5	Supply
S_1	35		20		10	65
S_2	5	20		30		55
S_3					50	50
Demand	40	20	20	30	60	170

20. Solve the transportation problem with three sources (with supplies 25, 25 and 50, respectively), four sinks (with demands 15, 20, 30 and 35, respectively) and the following cost matrix:

$$\mathbf{c} = \begin{bmatrix} 10 & 5 & 6 & 7 \\ 8 & 2 & 7 & 6 \\ 9 & 3 & 4 & 8 \end{bmatrix}$$

(This is the numerical problem discussed by Hitchcock, 1941.)

21. Solve the transportation problem with three sources (with supplies 50, 30 and 50, respectively), three sinks (with demands 60, 30 and 20, respectively) and the following cost matrix:

$$\mathbf{c} = \begin{bmatrix} 7 & 9 & 6 \\ - & 8 & 8 \\ 6 & - & 9 \end{bmatrix}$$

(The dashes in the matrix indicate the absence of the corresponding arc in the graph. It can be replaced by a large positive number M.)

22. Consider the transportation problem with two sources and three sinks. The supply at source 1 is 100 and the supply at source 2 is 80. The demand at sink 1 is p, where p is at least 40 and at most 100. The demand at sink 2 is q, where q is at least 50 and at most 140. The demand at sink 3 is $180 - (p + q)$. Explain how this problem (with inequality) constraints can be formulated as a transportation problem.

23. Solve the transportation problem with cost matrix

$$\mathbf{c} = \begin{bmatrix} 10 & 8 & 5 & 9 & 7 \\ 5 & M & 8 & 4 & 7 \\ 7 & 4 & 10 & 7 & 9 \\ 6 & 6 & 6 & 6 & 6 \end{bmatrix}$$

and supplies 20, 30, 30, 20 and demands 25, 25, 20, 10, 20.

24. Consider a 3×4 transportation problem with supplies 400, 200 and 300. The demands at sinks 1 and 2 are 200 and 300, respectively. The demand at sink 3 at least 100. Both sink 3 and sink 4 will consume any unused part of the total supply. Explain the procedure for solving this transportation problem.

25. Express the following doubly stochastic matrix $\mathbf{A}$ as a convex combination of permutation matrices:

$$\mathbf{A} = \begin{bmatrix} 0.21 & 0.13 & 0.38 & 0 & 0.28 \\ 0.33 & 0 & 0.15 & 0.38 & 0.14 \\ 0 & 0.73 & 0.14 & 0.05 & 0.08 \\ 0.38 & 0.14 & 0.05 & 0.36 & 0.07 \\ 0.08 & 0 & 0.28 & 0.21 & 0.43 \end{bmatrix}$$

26. Illustrate Konig's theorem on the bipartite graph (V, W, E), where $V = \{1, 2, 3, 4, 5\}$, $W = \{6, 7, 8, 9\}$ and E is the set $\{\{1, 6\}, \{2, 6\}, \{3, 7\}, \{3, 9\}, \{4, 6\}, \{5, 7\}, \{5, 8\}\}$.

27. Illustrate the Konig–Egervary theorem for the binary matrix defined by the bipartite graph in Exercise 26 above.

28. (Generalization of the Birkhoff–von Neuman theorem.) The sum of each line in a nonnegative $n \times n$ matrix $\mathbf{A}$ is M if and only if there exists positive numbers c_i $(i = 1, 2, \ldots, k)$ such that $A = c_1 \mathbf{P}_1 + c_2 \mathbf{P}_2 + \cdots + c_k \mathbf{P}_k$ and $c_1 + c_2 + \cdots + c_k = M$ where $\mathbf{P}_i$ are the permutations of the $n \times n$ identity matrix.

29. Show that the sum of each line in a binary square matrix $\mathbf{A}$ is M if and only if $\mathbf{A}$ is the sum of M permutation matrices.

30.
   (i) Verify that Hall's marriage condition is satisfied for the family $\{A_1, A_2, A_3, A_4, A_5, A_6\}$ where $A_1 = \{a, c\}$, $A_2 = \{b, c\}$, $A_3 = \{a, c, d, e\}$, $A_4 = \{b, d, e, f\}$, $A_5 = \{a, e\}$ and $A_6 = \{a, b\}$.
   (ii) Verify that Hall's marriage condition is not satisfied for the family $\{A_1, A_2, A_3, A_4, A_5, A_6\}$ where $A_1 = \{a, b, c\}$, $A_2 = \{b, c\}$, $A_3 = \{c, e, f\}$, $A_4 = \{a, b\}$, $A_5 = \{a, c\}$ and $A_6 = \{d, e, f\}$.

31. Let $X_1 = \{x_1, x_2, \ldots, x_n\}$ and $A_i = X - \{x_i\}$ for each i. Show that the family $\{A_i : i = 1, 2, \ldots, n\}$ has an SDR.

32. Obtain necessary and sufficient conditions for a family of sets to have a unique SDR.

33. Let $\{A_1, A_2, \ldots, A_n\}$ and $\{B_1, B_2, \ldots, B_n\}$ be two families of nonempty sets. Show that these two families will have a common SDR with n elements if and only if the following inequality holds for any subsets I and J of $\{1, 2, \ldots, n\}$:

$$|(\cup A_i : i \in I) \cap (\cup B_i : i \in J)| \geqslant |I| + |J| - n$$

(Notice that if each B_i is the union of all the sets B_i, the condition given here is precisely the marriage condition. So this condition is indeed a generalization of the marriage condition.)

34. Solve the following optimal assignment (minimization) problem:

$$A = \begin{bmatrix} 8 & 3 & 2 & 10 & 5 \\ 10 & 7 & 10 & 6 & 6 \\ 4 & 9 & 4 & 2 & 9 \\ 8 & 10 & 5 & 3 & 3 \\ 9 & 5 & 8 & 5 & 9 \end{bmatrix}$$

35. Solve the following optimal assignment (maximization) problem:

$$A = \begin{bmatrix} 22 & 19 & 20 & 13 & 18 \\ 16 & 16 & 12 & 12 & 19 \\ 16 & 18 & 20 & 21 & 19 \\ 16 & 13 & 14 & 11 & 11 \\ 21 & 11 & 20 & 19 & 22 \end{bmatrix}$$

36. Four candidates were tested by a firm wishing to fill three jobs. Candidate A scored 10, 9 and 4 points, respectively, when tested for these jobs. Candidate B scored 10, 6 and 8 points. Candidate C scored 9, 10 and 10 points. Candidate D scored 8, 9 and 8 points. The firm will hire three such that the sum of the scores of the three selected candidates is a maximum. Solve this assignment problem.

37. Solve the maximum weight bipartite matching problem for the bipartite graph $G = (V, W, E)$, where $V = \{1, 2, 3, 4, 5\}$ and $W = \{6, 7, 8, 9\}$. The edges $(1, 6)$ and $(1, 7)$ have weights 15 and 14, respectively. The edges $(2, 6)$ and $(2, 7)$ have weights 17 and 18. The edges $(3, 6)$, $(3, 7)$ and $(3, 8)$ have weights 12, 20 and 11. The edges $(4, 7)$, $(4, 8)$ and $(4, 9)$ have weights 12, 16 and 14. The edges $(5, 8)$ and $(5, 9)$ have weights 15 and 19. There are no other edges.

38. Solve the following optimal assignment (maximization) problem:

$$
\begin{bmatrix}
12 & 14 & 14 & 13 & 10 & 11 & 12 \\
14 & 16 & 16 & 15 & 14 & 13 & 15 \\
11 & 9 & 11 & 11 & 11 & 11 & 10 \\
11 & 13 & 13 & 13 & 12 & 11 & 13 \\
10 & 11 & 10 & 12 & 10 & 12 & 10 \\
15 & 17 & 17 & 14 & 17 & 16 & 17 \\
11 & 13 & 13 & 10 & 9 & 12 & 12
\end{bmatrix}
$$

3
Shortest path problems

3.1 SOME SHORTEST PATH ALGORITHMS

Statement of the problem

If v and w are two distinct vertices in a weighted digraph, a path of minimum weight (or minimum length) from v to w is called a **shortest path** (SP) **from v to w** and its length is called the **shortest distance** from v to w. If the graph is undirected, we define an SP and SD **between** two vertices.

The problem of finding such paths in weighted digraphs and graphs is known as the **shortest path problem**. In the case of an undirected connected graph, the SP problem is unbounded if there is an edge with negative weight. In the case of digraphs, if there is a directed cycle of C of negative weight then the problem of finding an SP from v to w becomes unbounded if there is a path from v to a vertex in C and a path from a vertex in C to w. So any algorithm to solve the SP problem in a directed graph should be capable of detecting a negative directed cycle in the digraph whenever it exists.

The shortest path problem as a transshipment problem

The SP problem can be viewed as a special case of transshipment and therefore we can adopt the procedure developed in Chapter 2 to solve any feasible SP problem. Suppose we are interested in finding the SP and SD from a vertex v to a vertex w, assuming that w is reachable from v. We now consider v as the only supply vertex with a total supply of 1 unit and w as the only demand vertex with a total demand of 1 unit. Every other vertex is an intermediate vertex. The weight (length) of the arc from i to j is the cost of sending one unit of the commodity from i to j along that arc. An optimal solution of this transshipment problem yields an SP from v to w and the corresponding SD which is the total cost. More generally, if there are n vertices in the digraph and if there is a path from v to every vertex, we can take v as the supply vertex with a total supply of $n-1$ units and every other vertex as a demand vertex with a demand of 1 unit. An optimal solution of this feasible transshipment problem readily yields an SP from v to every vertex in the network.

There are several algorithms for solving the SP problem other than the network simplex method. In this section we study two such algorithms. The first, due to Dijkstra (1959), is to find an SP and the SD from a specified vertex to every other vertex. We assume that the weight function is nonnegative. The second enables us to find an SP and the SD from every vertex to every other vertex in a digraph. If the graph is not strongly connected there will be at least one pair of vertices v and w such that w is not reachable from v, in which case the SP from v to w is a large positive value M. Quite appropriately, this is known as the **all-pairs shortest path problem**. The procedure to solve this problem is due to Floyd (1962) and Warshall (1962). The weight of an arc could be negative. But the algorithm will detect the existence of negative cycles and will terminate as soon as such a cycle is detected.

Dijkstra's algorithm

In the network $G = (V, E)$, let $V = \{1, 2, \ldots, n\}$ and let the weight of the arc (i, j) be $a(i, j)$, which is assumed to be nonnegative. If there is no arc from i to j, where i and j are two distinct vertices, then $a(i, j)$ is taken as a large positive number M. We thus have the $n \times n$ weight matrix $\mathbf{A} = [a(i, j)]$ in which the diagonal consists of n zeros. The optimization problem is to find the SD and SP from vertex 1 to all other vertices.

Define G and V as above. Each vertex i is assigned a label which is either permanent or tentative. The permanent label $L(i)$ of i is the SD from 1 to i, while the tentative label $L'(i)$ of i is an upper bound for the SD from 1 to i. At each stage of the procedure, P is the set of vertices with permanent labels and T is its complement. Initially $P = \{1\}$, with $L(1) = 0$ and $L'(i) = a(1, i)$ for each i.

The procedure is then as follows. Locate the vertex k in T such that $L'(k)$ is the smallest number in $\{L'(i): i \in T\}$. If there is a tie, then it is to be broken arbitrarily. Then the vertex k is assigned the permanent label $L(k)$ which is the same as $L'(k)$. The labels of the remaining vertices j in $T - \{k\}$ are now revised by replacing $L'(j)$ with the minimum of $L'(j)$ and $L(k) + a(k, j)$. The procedure terminates when $P = V$.

Theorem 3.1
Dijkstra's algorithm finds the SD from vertex 1 to vertex i for each i.

Proof
The proof is by induction on the cardinality of P. The induction hypothesis is that (i) $L(i)$ is equal to the SD from 1 to i for every i in P, and (ii) and for every j not in P, $L'(j)$ is the length of an SP from 1 to j, every intermediate vertex of which is in P.

Statement (i) is true when P has 1 element. Suppose it is true when P has $k-1$ vertices labeled $1, 2, \ldots, k-1$. By the induction hypothesis, just before vertex k is adjoined to the set P, $L'(k)$ is equal to the length of an SP from 1 to k in which every intermediate vertex is a vertex in P. Now k is adjoined to P, in which case P has k vertices.

Also $L'(k) = L(k)$. We claim $L(k)$ is the SD from 1 to k. If not, let d be the SD from 1 to k, implying that $d < L(k)$. So any SP from 1 to k should have at least one vertex not from P as an intermediate vertex.

Let v be the first such vertex from T in an SP from 1 to k. Suppose the SD from 1 to v is d'. Obviously $d' \leqslant d$. Now $d < L(k)$ which implies $d' < L(k)$. Thus there is a vertex v in T such that the SD from 1 to v is less than the SD from 1 to k. This violates the minimality requirement in choosing k from T. Thus it is true for k as well.

We now prove (ii) as follows. The vertex k is now used to revise the labels of the remaining $n - k$ vertices in T. The label $L'(j)$ either remains unchanged or is assigned the new label $L(k) + a(k, j)$. Thus the second inductive hypothesis holds when P has k vertices.

Each iteration in Dijkstra's algorithm essentially consists of two steps as follows.

Step 1 (designation of a permanent label). Find a vertex k in T for which $L'(k)$ is minimum. Tie-breaking is arbitrary. Declare k to be permanently labeled and adjoin k to the set P. Stop if $P = V$. Label the arc (i, k) where i is a labeled vertex that determines the minimum value of $L'(k)$.

Step 2 (revision of tentative labels). Replace $L'(j)$ by the smaller value of $L'(j)$ and $L(k) + a(k, j)$ for every j in T. Go to step 1.

The set of $n-1$ labeled arcs is acyclic since no arc is labeled if both its endpoints have a labeled arc incident to them. Thus the set of labeled arcs constitute a **shortest distance arborescence** rooted at vertex 1 in the network, giving both the SD and SP from 1 to each vertex.

Example 3.1
Obtain the SD from 1 to the remaining vertices in the directed network shown below, using Dijkstra's algorithm.

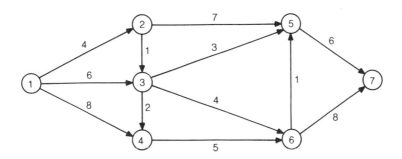

Iteration 1

Step 1.

$P = \{1\}$
$L(1) = 0$

$T = \{2, 3, 4, 5, 6, 7\}$
$L'(2) = 4, L'(3) = 6, L'(4) = 8$
$L'(5) = L'(6) = L'(7) = M$
Vertex 2 is assigned a permanent label.
Arc (1, 2) is labeled.

Step 2.

$P = \{1, 2\}$
$L(2) = 4$

$T = \{3, 4, 5, 6, 7\}$
$L'(3) = \min \{6, L(2) + a(2, 3)\}$
$L'(4) = \min \{8, L(2) + a(2, 4)\}$
$L'(5) = \min \{M, L(2) + a(2, 5)\}$
$L'(6) = \min \{M, L(2) + a(2, 6)\}$
$L'(7) = \min \{M, L(2) + a(2, 7)\}$

Iteration 2

Step 1.

$P = \{1, 2\}$
$L(1) = 0$
$L(2) = 4$

$T = \{3, 4, 5, 6, 7\}$
$L'(3) = 5, L'(4) = 8, L'(5) = 11$
$L'(6) = L'(7) = M$
Vertex 3 is assigned a permanent label.
Arc (2, 3) is labeled.

Step 2.

$P = \{1, 2, 3\}$
$L(3) = 5$

$T = \{4, 5, 6, 7\}$
$L'(4) = \min \{8, L(3) + a(3, 4)\}$
$L'(5) = \min \{11, L(3) + a(3, 5)\}$
$L'(6) = \min \{M, L(3) + a(3, 6)\}$
$L'(7) = \min \{M, L(3) + a(3, 7)\}$

Iteration 3

Step 1.

$P = \{1, 2, 3\}$
$L(1) = 0$
$L(2) = 4$
$L(3) = 5$

$T = \{4, 5, 6, 7\}$
$L'(4) = 7, L'(5) = 10,$
$L'(6) = 9, L'(7) = M$
Vertex 4 is assigned a permanent label.
Arc (3, 4) is labeled.

Step 2.

$P = \{1, 2, 3, 4\}$
$L(4) = 7$

$T = \{5, 6, 7\}$
$L'(5) = \min \{10, L(4) + a(4, 5)\}$
$L'(6) = \min \{9, L(4) + a(4, 6)\}$
$L'(7) = \min \{M, L(4) + a(4, 7)\}$

Iteration 4

Step 1.

$P = \{1, 2, 3, 4\}$
$L(1) = 0, L(2) = 4,$
$L(3) = 5, L(4) = 7$

$T = \{5, 6, 7\}$
$L'(5) = 10, L'(6) = 9,$
$L'(7) = M$
Vertex 6 is assigned a permanent label.
Arc (3, 6) is labeled.

Step 2.

$P = \{1, 2, 3, 4, 6\}$
$L(6) = 9$

$T = \{5, 7\}$
$L'(5) = \min \{10, L(6) + a(6, 5)\}$
$L'(7) = \min \{M, L(6) + a(6, 7)\}$

Iteration 5

Step 1.

$P = \{1, 2, 3, 4, 6\}$
$L(1) = 0, L(2) = 4,$
$L(3) = 5, L(4) = 7,$
$L(6) = 9$

$T = \{5, 7\}$
$L'(5) = 10, L'(7) = 17$
Vertex 5 is assigned a permanent label.
Arc (3, 5) is labeled.

Step 2.

$P = \{1, 2, 3, 4, 6, 5\}$
$L(5) = 10$

$T = \{7\}$
$L'(7) = \min \{17, L(5) + a(5, 7)\}$

Iteration 6

Step 1.

$P = \{1, 2, 3, 4, 6, 5\}$
$L(1) = 0, L(2) = 4,$
$L(3) = 5, L(4) = 7,$
$L(6) = 9, L(5) = 10.$

$T = \{7\}$
$L'(7) = 16$
Vertex 7 gets a permanent label.
Arc (5, 7) is labeled.

Step 2.

$P = \{1, 2, 3, 4, 6, 5, 7\}$
$L(7) = 16.$

T is empty

Thus $L(1) = 0, L(2) = 4, L(3) = 5, L(4) = 7, L(6) = 9, L(5) = 10$ and $L(7) = 16$, giving the SD from 1 to each vertex. The labeled arcs are (1, 2), (2, 3), (3, 4), (3, 5), (3, 6) and (5, 7) which constitute a shortest distance arborescence in the given network as shown below, giving the SP from vertex 1 to every other vertex.

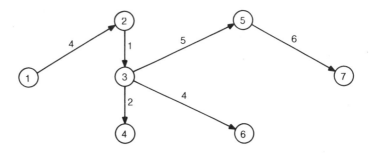

Note that Dijkstra's algorithm need not solve the SD problem for an arbitrary weight function. Consider the network $G = (V, E)$ where $V = \{1, 2, 3\}$ and the arcs are $(1, 2), (1, 3)$ and $(2, 3)$ with weights $10, 8$ and -3, respectively. In iteration 2 we obtain $L(3) = 8$ but the SD from 1 to 3 is only 7.

The Floyd–Warshall algorithm

Consider a directed network $G = (V, A)$ with $V = \{1, 2, \ldots, n\}$ and with an arbitrary weight function. Let $\mathbf{A} = [a_{uv}]$ be the $n \times n$ weight matrix and let $\mathbf{P} = [p_{uv}]$ be the $n \times n$ matrix where $p_{uv} = v$.

There are n iterations during the execution of the algorithm. Iteration j begins with two $n \times n$ matrices $\mathbf{A}(j-1)$ and $\mathbf{P}(j-1)$ and ends with $\mathbf{A}(j)$ and $\mathbf{P}(j)$. Initially, $\mathbf{A}(0) = \mathbf{A}$ and $\mathbf{P}(0) = \mathbf{P}$. The (u, v)th elements in $\mathbf{A}(j)$ and $\mathbf{B}(j)$ are denoted $a_{uv}(j)$ and $p_{uv}(j)$, respectively.

For a fixed j, the matrices $\mathbf{A}(j)$ and $\mathbf{P}(j)$ are obtained from the matrices $\mathbf{A}(j-1)$ and $\mathbf{P}(j-1)$ by applying the following rule known as the **triangle operation**:

1. If $a_{uv}(j-1) \leqslant a_{uj}(j-1) + a_{jv}(j-1)$ then $a_{uv}(j) = a_{uv}(j-1)$ and $p_{uv}(j) = p_{uv}(j-1)$.
2. Otherwise $a_{uv}(j) = a_{uj}(j-1) + a_{jv}(j-1)$. In this case $p_{uv}(j) = p_{uj}(j-1)$.

When the algorithm terminates we are left with two matrices $\mathbf{A}(n)$ and $\mathbf{P}(n)$. We now prove that the former gives the SD and the latter gives an SP from any vertex u to any vertex w.

Theorem 3.2
If we perform the triangle operation successively for the vertices $j = 1, 2, \ldots, n$, then $a_{uv}(n)$, the (u, v)th entry in $\mathbf{A}(n)$, becomes equal to the SD from u to v.

Proof
Assume that the weight function is nonnegative. The proof is by induction on j. The induction hypothesis is that when the triangle operation is performed for $j = j_0$, $a_{uv}(j_0)$ is the SD from u to v using a path with intermediate vertices $w \leqslant j_0$. This is true when $j_0 = 1$. Suppose it is true for $j = j_0 - 1$.

Consider the triangle operation for $j = j_0$ which determines the value of $a_{uv}(j_0)$. If the shortest path from u to v with intermediate vertices $w \leq j_0$ does not pass through j_0, then $a_{uv}(j_0)$ will be the same as $a_{uv}(j_0 - 1)$, in which case the inductive hypothesis holds for $j = j_0$.

On the other hand, if the shortest path from u to v with intermediate vertices $w \leq j_0$ passes through the vertex j, then $a_{uv}(j_0)$ will be the sum of $a_{uj}(j_0 - 1)$ and $a_{jv}(j_0 - 1)$. But, by the induction hypothesis, each of these entries corresponds to an optimal path with intermediate vertices $w \leq j_0 - 1$. So the sum of the two entries also corresponds to an optimal path with intermediate vertices $w \leq j_0$. So in this case also, it is true when $j = j_0$. This completes the inductive argument with the nonnegativity restriction.

Now we relax the nonnegativity assumption. The procedure proved above still works as long as there are no negative directed cycles in the network. If there is a negative directed cycle, at some stage a diagonal element of $A(j)$ will become negative and the algorithm will terminate at that stage.

If $p_{uv}(n) = w$ then an SP from u to v is obtained by joining (concatenating) the SP from u to w and the SP from w to v. This is a consequence of the triangle operation. By working backwards using the elements in the matrix $P(n)$ a shortest path from u to v can be identified. This completes the proof of the theorem.

Example 3.2
Obtain the SD matrix and the SP matrix in the directed network as shown below.

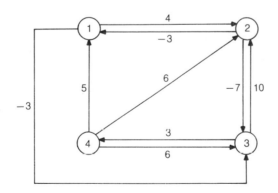

We begin with the following matrices:

$$A = A(0) = \begin{bmatrix} 0 & 4 & -3 & M \\ -3 & 0 & -7 & M \\ M & 10 & 0 & 3 \\ 5 & 6 & 6 & 0 \end{bmatrix} \quad \text{and } P = P(0) = \begin{bmatrix} 1 & 2 & 3 & 4 \\ 1 & 2 & 3 & 4 \\ 1 & 2 & 3 & 4 \\ 1 & 2 & 3 & 4 \end{bmatrix}$$

Iteration 1, **based at vertex 1 ($j = 1$)**

$$a_{11}(1) = 0 \qquad\qquad\qquad\qquad\qquad\qquad\qquad\qquad\qquad\qquad p_{11}(1) = 1$$
$$a_{12}(1) = \min\{a_{12}(0), a_{11}(0) + a_{12}(0)\} = 4 \text{ (no change)} \qquad p_{12}(1) = 2$$
$$a_{13}(1) = \min\{a_{13}(0), a_{11}(0) + a_{13}(0)\} = -3 \text{ (no change)} \quad p_{13}(1) = 3$$
$$a_{14}(1) = \min\{a_{14}(0), a_{11}(0) + a_{14}(0)\} = M \text{ (no change)} \qquad p_{14}(1) = 4$$
$$a_{21}(1) = \min\{a_{21}(0), a_{21}(0) + a_{11}(0)\} = -3 \text{ (no change)} \quad p_{21}(1) = 1$$
$$a_{22}(1) = \min\{a_{22}(0), a_{21}(0) + a_{12}(0)\} = 0 \text{ (no change)} \qquad p_{22}(1) = 2$$
$$a_{23}(1) = \min\{a_{23}(0), a_{21}(0) + a_{13}(0)\} = -7 \text{ (no change)} \quad p_{23}(1) = 3$$
$$a_{24}(1) = \min\{a_{24}(0), a_{21}(0) + a_{14}(0)\} = M \text{ (no change)} \qquad p_{24}(1) = 4$$
$$a_{31}(1) = \min\{a_{31}(0), a_{31}(0) + a_{11}(0)\} = M \text{ (no change)} \qquad p_{31}(1) = 1$$
$$a_{32}(1) = \min\{a_{32}(0), a_{31}(0) + a_{12}(0)\} = 10 \text{ (no change)} \quad p_{32}(1) = 2$$
$$a_{33}(1) = \min\{a_{33}(0), a_{31}(0) + a_{13}(0)\} = 0 \text{ (no change)} \qquad p_{33}(1) = 3$$
$$a_{34}(1) = \min\{a_{34}(0), a_{31}(0) + a_{14}(0)\} = 3 \text{ (no change)} \qquad p_{34}(1) = 4$$
$$a_{41}(1) = \min\{a_{41}(0), a_{41}(0) + a_{11}(0)\} = 5 \text{ (no change)} \qquad p_{41}(1) = 1$$
$$a_{42}(1) = \min\{a_{42}(0), a_{41}(0) + a_{12}(0)\} = 6 \text{ (no change)} \qquad p_{42}(1) = 2$$
$$a_{43}(1) = \min\{a_{43}(0), a_{41}(0) + a_{13}(0)\} = 2 \text{ (\textit{change})} \qquad\quad p_{43}(1) = 1$$
$$a_{44}(1) = \min\{a_{44}(0), a_{41}(0) + a_{14}(0)\} = 0 \text{ (no change)} \qquad p_{44}(1) = 4$$

Thus at the end of the first iteration we have the following matrices:

$$\mathbf{A}(1) = \begin{bmatrix} 0 & 4 & -3 & M \\ -3 & 0 & -7 & M \\ M & 10 & 0 & 3 \\ 5 & 6 & 2 & 0 \end{bmatrix} \quad \text{and} \quad \mathbf{P}(1) = \begin{bmatrix} 1 & 2 & 3 & 4 \\ 1 & 2 & 3 & 4 \\ 1 & 2 & 3 & 4 \\ 1 & 2 & 1 & 4 \end{bmatrix}$$

Iteration 2, **based at vertex 2 ($j = 2$)**

This begins with $\mathbf{A}(1)$ and $\mathbf{P}(1)$. The triangle operations are carried out as in the previous operation – at this stage the $a_{uv}(2) = \min\{a_{uv}(1), a_{u2}(1) + a_{2v}(1)\}$. At the end of this iteration we have the following matrices:

$$\mathbf{A}(2) = \begin{bmatrix} 0 & 4 & -3 & M \\ -3 & 0 & -7 & M \\ 7 & 10 & 0 & 3 \\ 3 & 6 & -1 & 0 \end{bmatrix} \quad \text{and} \quad \mathbf{P}(2) = \begin{bmatrix} 1 & 2 & 3 & 4 \\ 1 & 2 & 3 & 4 \\ 2 & 2 & 3 & 4 \\ 2 & 2 & 2 & 4 \end{bmatrix}$$

Iteration 3, **based at vertex 3 ($j = 3$)**

This begins with $\mathbf{A}(2)$ and $\mathbf{P}(2)$. By applying the triple operation, we obtain the following matrices:

$$\mathbf{A}(3) = \begin{bmatrix} 0 & 4 & -3 & 0 \\ -3 & 0 & -7 & -4 \\ 7 & 10 & 0 & 3 \\ 3 & 6 & -1 & 0 \end{bmatrix} \quad \text{and} \quad \mathbf{P}(3) = \begin{bmatrix} 1 & 2 & 3 & 3 \\ 1 & 2 & 3 & 3 \\ 2 & 2 & 3 & 4 \\ 2 & 2 & 2 & 4 \end{bmatrix}$$

Iteration 4, **based at vertex 4 ($j=4$)**
This begins with A(3) and P(3). By applying the triple operation, we get the following matrices:

$$\mathbf{A}(4) = \begin{bmatrix} 0 & 4 & -3 & 0 \\ -3 & 0 & -7 & -4 \\ 6 & 9 & 0 & 3 \\ 3 & 6 & -1 & 0 \end{bmatrix} \quad \text{and } \mathbf{P}(4) = \begin{bmatrix} 1 & 2 & 3 & 3 \\ 1 & 2 & 3 & 3 \\ 4 & 4 & 3 & 4 \\ 2 & 2 & 2 & 4 \end{bmatrix}$$

At this stage we have the SD from every vertex to every other vertex in the network, which can be obtained readily from A(4).

The various shortest paths are obtained as follows:

- From 1 to 2: $p_{12}(4) = 2$. So the SP is $1\rightarrow2$, which is the arc (1, 2).
- From 1 to 3: $p_{13}(4) = 3$. So the SP from 1 to 3 is the arc (1, 3).
- From 1 to 4: $p_{14}(4) = 3$. So take the arc (1, 3) and join it to the SP from 3 to 4. We have $p_{34}(4) = 4$. So the SP from 3 to 4 is the arc (3, 4). Thus the SP is $1\rightarrow3\rightarrow4$.
- From 2 to 1: the SP is the arc (2, 1).
- From 2 to 3: the SP is the arc (2, 3).
- From 2 to 4: the SP is $2\rightarrow3\rightarrow4$.
- From 3 to 1: the SP is $3\rightarrow4\rightarrow2\rightarrow1$.
- From 3 to 2: the SP is $3\rightarrow4\rightarrow2$.
- From 3 to 4: the SP is the arc (3, 4).
- From 4 to 1: the SP is $4\rightarrow2\rightarrow1$.
- From 4 to 2: the SP is the arc (4, 2).
- From 4 to 3: the SP is $4\rightarrow2\rightarrow3$.

Locating negative cycles

Consider a network for which the weight matrix is as follows:

$$\mathbf{A} = \begin{bmatrix} 0 & M & M & 1 \\ 2 & 0 & 1 & 3 \\ M & M & 0 & M \\ 2 & -4 & 3 & 0 \end{bmatrix}$$

At the end of the second iteration we have the following matrices:

$$\mathbf{A}(2) = \begin{bmatrix} 0 & M & M & 1 \\ 2 & 0 & 1 & 3 \\ M & M & 0 & M \\ 2 & -4 & -3 & -1 \end{bmatrix} \quad \text{and } \mathbf{P}(2) = \begin{bmatrix} 1 & 2 & 3 & 4 \\ 1 & 2 & 3 & 1 \\ 1 & 2 & 3 & 4 \\ 2 & 2 & 2 & 2 \end{bmatrix}$$

In **A**(2), the diagonal element (4, 4) is negative, indicating the presence of a negative cycle in the network. The **P**(2) matrix locates the closed simple path from 4 to 4 as the negative cycle $4 \to 2 \to 1 \to 4$.

Observe that in the case of Dijkstra's algorithm, at each iteration one vertex from the set T (or temporarily labeled vertices) is assigned a permanent label and then is adjoined to the set P of permanently labeled vertices. But in the case of the Floyd–Warshall algorithm, the entries in the matrix **A**(j) are assigned permanent values at the end of the iteration. So even though both procedures are iterative, assigning a temporary estimate which is an upper bound of the SD, there is a subtle distinction between them which in a way classifies SP problems into two groups: problems (with nonnegative weight functions) in which the labels become permanent once they are assigned; and problems (with arbitrary weight functions) in which assigned labels become permanent at the end of the algorithm. In this section we have analyzed a prototype problem from each category.

3.2 BRANCH AND BOUND METHODS FOR SOLVING THE TRAVELING SALESMAN PROBLEM

The most famous of all network optimization problems is the celebrated traveling salesman problem (TSP) which, as the single most important 'unsolved' combinatorial optimization problem, plays a significant role in the development of optimization concepts and analysis of algorithms. The problem is unsolved in the sense that so far nobody has developed an algorithm that satisfies any 'formal or informal standards of efficiency' (Hoffman and Wolfe, 1985).

A Hamiltonian cycle is a (directed) cycle in a (directed) graph which starts at a vertex and ends at the same vertex after passing through every other vertex exactly once. The problem of obtaining a Hamiltonian cycle (if it exists) with weight as small as possible is the **optimal Hamiltonian problem** (OHP).

In many practical situations a more meaningful question to ask is whether there exists a closed path which passes through each vertex at least once. The problem of finding a closed path of this kind with minimum weight is the **optimal salesman problem** (OSP). It is possible to have a salesman's route in a network G even when G is not Hamiltonian.

Suppose **A** is the weight (distance) matrix of a network $G = (V, E)$ and **D** is its shortest distance (SD) matrix which will give the SD between pairs of vertices. Let $G' = (V, E')$ be the network in which the length of the arc from i to j is the SD from i to j in G. In **A**, the (i, j)th entry is a large positive number M if (i, j) is not an arc in G. In G', the (i, j)th entry is M if j is not reachable from i. For obvious reasons, the diagonal elements in both **A** and **D** are also assigned the value M. The matrix **A** is used to solve the OHP, whereas the OSP is solved using the matrix **D**. Any algorithm which can be used to obtain an optimal Hamiltonian cycle in G can be applied to obtain such a cycle in G' from which an optimal salesman's circuit can be constructed.

A path between two vertices in a connected undirected graph G is a Hamiltonian path if it passes through each vertex exactly once. A Hamiltonian path in G is therefore a spanning tree in which the degree of each vertex is at most 2. Conversely, a spanning tree in G is a Hamiltonian path if the degree of each vertex is at most 2. So the problem of finding a Hamiltonian path with minimum weight is equivalent to the problem of finding a minimum spanning tree in which the degree of each vertex is at most 2. Thus the MST problem is a relaxation of the optimal Hamiltonian path problem, and the latter is a restriction of the former.

Conversely, the problem of finding a Hamiltonian cycle in G starting from vertex 1 and ending in vertex 1 in $G = (V, E)$, where $V = \{1, 2, \ldots, n\}$, is equivalent to the problem of finding a Hamiltonian path from 1 to $n+1$ in $G' = (V', E')$, where V' is the set obtained by adjoining a new vertex $n+1$ to the set V and E' consists of all edges in E together with edges of the form $\{i, n+1\}$ whenever the edge $\{1, i\}$ is in E. The MST problem is therefore a relaxation of the OHP as well.

Next we relate the optimal assignment problem (section 2.6) to the OHP. Corresponding to the undirected graph $G = (V, E)$ with $V = \{1, 2, \ldots, n\}$ we construct a bipartite graph $G' = (S, T, F)$ where $S = \{s_i : i = 1, 2, \ldots, n\}$ and $T = \{t_j : j = 1, 2, \ldots, n\}$, and join s_i and t_j by an edge f if and only if $e = \{i, j\}$ is in E. Then a Hamiltonian cycle in G obviously corresponds to a unique assignment (complete matching) in G. Thus the optimal assignment problem is also a relaxation of the OHP.

Now if an optimal assignment in G' corresponds to a cycle C in G, then C is an optimal Hamiltonian cycle in G. But an optimal assignment in G' need not correspond to a cycle. For example, the matching $M = \{\{s_1, t_2\}, \{s_2, t_1\}, \{s_3, t_4\}, \{s_4, t_3\}\}$ in G' corresponds not to a Hamiltonian cycle (tour) in G but to two cycles (subtours), namely, the cycle 1—2—1 and the cycle 3—4—3. As a matter of fact it is immaterial whether G is directed or not. Thus the OHP in G can be reformulated as the following problem. Given G, construct G' and then find an optimal assignment in G' with no subtours if such an assignment exists. This is easier said than done. What creates the combinatorial explosion is the fact that the number of subtours could be large even when the number of vertices is not very large.

Optimal assignment relaxation

We first discuss a procedure to solve the optimal Hamiltonian problem by subtour elimination using the optimal assignment relaxation and a technique known as **branch and bound**. Our discussion of the algorithm is based on Bellmore and Malone (1971).

Let $\mathbf{A}$ be the $n \times n$ weight matrix for G and let Z^* be the weight of an optimal Hamiltonian cycle in G. (Z^* is a large positive value M if there is no

Hamiltonian cycle.) Let $w(\mathbf{A})$ be the weight of the optimal assignment on $\mathbf{A}$. Then $w(\mathbf{A}) \leqslant Z^*$. Thus $w(\mathbf{A})$ is a lower bound for Z^*.

Suppose the optimal assignment on $\mathbf{A}$ gives a subtour denoted by $v_1 \to v_2 \to \cdots \to v_k \to v_1$, where $k < n$. Let S be the set of these k vertices in the subtour. If V is partitioned into two subsets S and S', any Hamiltonian cycle in G should contain at least one arc from a vertex in S to a vertex in S'.

So we create k subproblems corresponding to the k vertices in S. For the ith subproblem, there are no arcs from v_i to the other arcs in S. So in the weight matrix $\mathbf{A}$, the weight of an arc from a vertex in S to another vertex in S is assigned the value M and then we have the weight matrix $\mathbf{A}_i$ corresponding to the vertex v_i. We thus have k **branching matrices**.

We now solve $\mathbf{A}_i$ as an optimal assignment problem for each vertex in S. Each set will give a solution. Since any Hamiltonian cycle in G should contain at least one arc from a vertex out of these k vertices in S to a vertex in S', it is enough if we examine only these k optimal solutions for any future computation in our attempt to obtain an optimal Hamiltonian cycle. No Hamiltonian cycles are lost in this branching process. At the same time, the subtour created by the vertices in S is eliminated.

Usually the set S which constitutes a subtour is chosen for branching such that S has as few elements as possible. This need not be the best choice even though it is intuitively appealing. If any of these subproblems gives a Hamiltonian cycle and if its weight is less than or equal to the lower bounds on the other subproblems, then that cycle is optimal. Otherwise, we take that subproblem with the lowest bound and branch again according to the subtours in the subproblem. This iterative process ultimately leads us to an optimal solution.

Thus the 'branch and bound' procedure described above involves three steps:

Step 1. Solve the associated assignment problem. If there are no subtours, the current solution is optimal. Otherwise go to step 2.

Step 2. Take the set of vertices in a subtour and branch the problem into subproblems. Solve each subproblem as an optimal assignment problem. Go to step 3.

Step 3. The solutions without subtours are compared with the best existing solution. The solution with the minimum weight becomes the current best solution. The solutions with subtours, with weight less than the current best solution, are considered. The subproblem with the minimum value is located. If the weight of this solution is more than the current best solution, then the current best solution is optimal. Otherwise, go to step 2.

Example 3.3

Using the branch and bound method described above, find an optimal Hamiltonian cycle and an optimal salesman tour in the network shown below, with weight matrix **A** given by

$$\mathbf{A} = \begin{bmatrix} M & 17 & 10 & 15 & 17 \\ 18 & M & 6 & 10 & 20 \\ 12 & 5 & M & 14 & 19 \\ 12 & 11 & 15 & M & 7 \\ 16 & 21 & 18 & 6 & M \end{bmatrix}$$

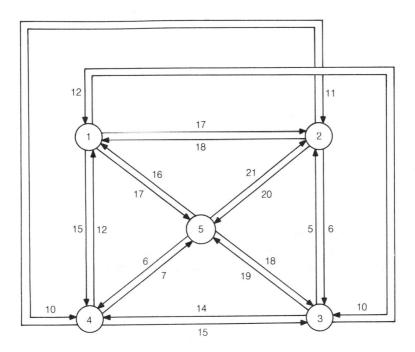

Optimal Hamiltonian problem

The vertices are 1, 2, 3, 4 and 5. By solving **A** as an assignment problem, we get two subtours $1 \to 3 \to 2 \to 1$ and $4 \to 5 \to 4$ with $w(\mathbf{A}) = 46$.

Thus 46 is a lower bound for the weight Z^* of an optimal Hamiltonian cycle in G. The current solution with weight 46 is non-Hamiltonian with weight $w(\mathbf{A}) = 46$. So we write $w(\mathbf{A}) = 46$ (NH). We take $S = \{4, 5\}$ and $S' = \{1, 2, 3\}$. Then the two branching matrices are

$$\mathbf{A}(4) = \begin{bmatrix} M & 17 & 10 & 15 & 17 \\ 18 & M & 6 & 10 & 20 \\ 12 & 5 & M & 14 & 19 \\ 12 & 11 & 15 & M & M \\ 16 & 21 & 18 & 6 & M \end{bmatrix} \text{ and } \mathbf{A}(5) = \begin{bmatrix} M & 17 & 10 & 15 & 17 \\ 18 & M & 6 & 10 & 20 \\ 12 & 5 & M & 14 & 19 \\ 12 & 11 & 15 & M & 7 \\ 16 & 21 & 18 & M & M \end{bmatrix}$$

The optimal assignment on $\mathbf{A}(4)$ gives two subtours, $1 \to 5 \to 4 \to 1$ and $2 \to 3 \to 2$, with weight 46.

The optimal assignment on $\mathbf{A}(5)$ gives the Hamiltonian cycle $1 \to 3 \to 2 \to 4 \to 5 \to 1$, with weight 48.

Thus $w(\mathbf{A}(4)) = 46$ (NH) and $w(\mathbf{A}(5)) = 48$ (H).

The current best solution is at $\mathbf{A}(5)$ and $Z^* \leqslant 48$. In $\mathbf{A}(4)$, the weight is less than the weight of the current best solution. So we branch out from $\mathbf{A}(4)$. For this branching, $S = \{2, 3\}$ and $S' = \{1, 4, 5\}$.

The two branching matrices coming out of $\mathbf{A}(4)$ are denoted by $\mathbf{A}(4, 2)$ and $\mathbf{A}(4, 3)$. $\mathbf{A}(4, 2)$ is obtained from $\mathbf{A}(4)$ by making the $(2, 3)$th element take the value M. $\mathbf{A}(4, 3)$ is obtained by making the $(3, 2)$th element in $\mathbf{A}(4)$ take the value M.

We now solve both $\mathbf{A}(4, 2)$ and $\mathbf{A}(4, 3)$ as optimal assignment problems.

In $\mathbf{A}(4, 2)$, we have $1 \to 3 \to 2 \to 5 \to 4 \to 1$, with weight $w(\mathbf{A}(4, 2)) = 53$ (H).

In $\mathbf{A}(4, 3)$, we have $1 \to 5 \to 4 \to 2 \to 3 \to 1$, with weight $w(\mathbf{A}(4, 3)) = 52$ (H).

Thus we have a **tree enumeration scheme** (Fig. 3.1) in this branch and bound method depicted by an arborescence with root at the initial weight matrix $\mathbf{A}$ and vertices at subsequent branching matrices. From $\mathbf{A}$ there are two arcs – one from $\mathbf{A}$ to $\mathbf{A}(4)$ and the other from $\mathbf{A}$ to $\mathbf{A}(5)$. Since $\mathbf{A}(5)$ gives a Hamiltonian vertex, there is no branching from $\mathbf{A}(4)$ and so its outdegree is 0. Since the (non-Hamiltonian) weight at $\mathbf{A}(4)$ is less than the weight of the current best solution (namely the solution at $\mathbf{A}(5)$), we have to branch out from $\mathbf{A}(4)$ and thus the outdegree of $\mathbf{A}(4)$ is not 0. From $\mathbf{A}(4)$ there are two arcs – one to $\mathbf{A}(4, 2)$ and the other to $\mathbf{A}(4, 3)$. Each gives a Hamiltonian cycle. So the outdegree is 0 in each case. In the arborescence, a vertex with outdegree 0 with

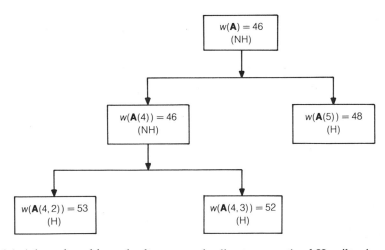

Fig. 3.1 A branch and bound arborescence leading to an optimal Hamiltonian cycle by using the optimal assignment of the cost matrix.

least weight corresponds to an optimal solution. An optimal solution is the cycle shown below.

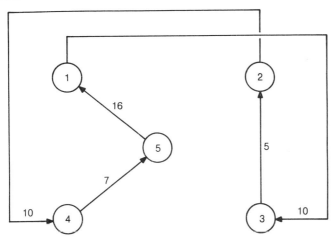

Optimal salesman problem

Now we take the SD matrix **D** as the initial matrix:

$$\mathbf{D} = \begin{bmatrix} M & 15 & 10 & 15 & 17 \\ 18 & M & 6 & 10 & 17 \\ 12 & 5 & M & 14 & 19 \\ 12 & 11 & 15 & M & 7 \\ 16 & 17 & 18 & 6 & M \end{bmatrix}$$

The optimal assignment on **D** gives two subtours, $1 \to 5 \to 4 \to 1$ and $2 \to 3 \to 2$, with $w(\mathbf{D}) = 46$ (NH). Let $S = \{2, 3\}$ and $S' = \{1, 3, 4\}$.

The optimal assignment on **D**(2) gives two subtours, as $1 \to 3 \to 2 \to 1$ and $4 \to 5 \to 4$, and $w(\mathbf{D}(2)) = 46$ (NH).

The optimal assignment on **D**(3) also gives two subtours, $1 \to 2 \to 3 \to 1$; $4 \to 5 \to 4$, and $w(\mathbf{D}(3)) = 46$ (NH).

At this stage we have to branch out from both **D**(2) and **D**(3).

Branching out from **D**(2), we have $S = \{4, 5\}$. The branching matrices are **D**(2, 4) and **D**(2, 5). The optimal assignments on these two matrices give Hamiltonian cycles $1 \to 3 \to 2 \to 5 \to 4 \to 1$, with weight $w(\mathbf{D}(2, 4)) = 50$, and $1 \to 3 \to 2 \to 4 \to 5 \to 1$, with weight $w(\mathbf{D}(2, 5)) = 48$. So both **D**(2, 4) and **D**(2, 5) are terminal vertices in the arborescence.

Branching out from **D**(3), we again have $S = \{4, 5\}$. The branching matrices are **D**(3, 4) and **D**(3, 5). The optimal assignment on **D**(3, 4) gives the Hamiltonian cycle $1 \to 5 \to 4 \to 2 \to 3 \to 1$, with $w(\mathbf{D}(3, 4)) = 52$. The optimal assignment on **D**(3, 5) gives subtours $1 \to 3 \to 1$ and $2 \to 4 \to 5 \to 2$, with a total weight $w(\mathbf{D}(3, 4)) = 56$. The weight of the current optimal solution is 48 and is less than the weight of the assignment on **D**(3, 4) which has subtours. So we do not branch out from **D**(3, 4).

The optimal solution is from the matrix **D**(2, 5) with weight 48.

In Example 3.3 the solution to the OSP is the same as the solution to the OHP. Next we consider an example in which the weight of the solution of the OSP is less than that of the OHP.

Example 3.4
Suppose the weight matrix is

$$\mathbf{A} = \begin{bmatrix} M & 5 & 19 & 11 \\ M & M & 4 & 7 \\ M & 5 & M & 14 \\ 9 & M & 6 & M \end{bmatrix}$$

corresponding to the network shown below:

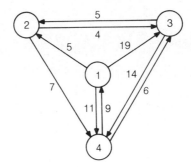

The solution to the OHP is $1 \to 2 \to 3 \to 4 \to 1$, with a total weight of 32:

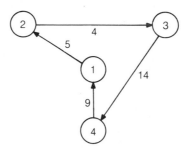

The shortest distance matrix for this problem is

$$\mathbf{D} = \begin{bmatrix} M & 5 & 9 & 11 \\ 16 & M & 4 & 7 \\ 21 & 5 & M & 12 \\ 9 & 11 & 6 & M \end{bmatrix}$$

The solution to the OHP using this matrix gives the closed path $1 \to 3 \to 2 \to 4 \to 1$, with a total weight of 30. In this closed path, each arc (i, j) is a shortest path from i to j. So we get the circuit $1 \to 2 \to 3 \to 2 \to 4 \to 1$, which is an optimal salesman tour with weight less than that of an optimal Hamiltonian cycle:

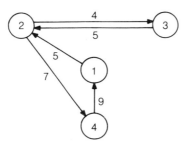

Weight matrix reduction

It is possible to use a branch and bound procedure to solve the optimal Hamiltonian problem without using the optimal assignment of the weight matrix. This method is due to Little *et al.* (1963). The algorithm is explained here by using the same numerical example as appeared in their paper.

Initialization

Suppose the cost of a Hamiltonian cycle C in a network is $Z(C)$. If a constant h is subtracted from each element in a row (or a column) of the cost matrix $\mathbf{A}$, resulting in a new matrix $\mathbf{A}'$, we say that the matrix $\mathbf{A}'$ is obtained from $\mathbf{A}$ by **reducing** a row (or a column). The cost of the cycle C under $\mathbf{A}'$ is $z(C) - h$. Thus a Hamiltonian cycle under $\mathbf{A}$ is optimal if and only if it is optimal under $\mathbf{A}'$.

We assume that the cost (weight) of each arc in the network under consideration is nonnegative. The matrix $\mathbf{A}'$ obtained from $\mathbf{A}$ by reducing rows and columns such that there is a zero in each row and in each column is the **reduced matrix** of $\mathbf{A}$. Let h be the sum of the nonnegative numbers used to obtain $\mathbf{A}'$ from $\mathbf{A}$. If $Z(C)$ and $Z'(C)$ are the costs of C under $\mathbf{A}$ and $\mathbf{A}'$, respectively, then it is easy to see that $Z'(C) + h = Z(C)$, which implies $h \leqslant Z(C)$ for any Hamiltonian cycle C in the network. Thus h is an **evaluation by lower bound** for the set Ω of all Hamiltonian cycles in G which is the root of the arborescence associated with this procedure.

For example, consider the six-city TSP for which the weight matrix is as follows:

$$\mathbf{A} = \begin{bmatrix} M & 27 & 43 & 16 & 30 & 26 \\ 7 & M & 16 & 1 & 30 & 25 \\ 20 & 13 & M & 35 & 5 & 0 \\ 21 & 16 & 25 & M & 18 & 18 \\ 12 & 46 & 27 & 28 & M & 5 \\ 23 & 5 & 5 & 9 & 5 & M \end{bmatrix}$$

The six cities are: a, b, c, d, e and f (represented by rows 1, 2, 3, 4, 5 and 6 of **A**, respectively).

We subtract 16 from row a, 1 from row b, 0 from row c, 16 from row d, 5 from row e and 5 from row f. Column a is still a nonzero column. So we subtract 5 from column a. Thus $h = (16 + 1 + 0 + 16 + 5 + 5) + 5 = 48$, which is the initial evaluation by lower bound at the root Ω of the arborescence. The initial reduced matrix is

$$\mathbf{A'} = \begin{bmatrix} M & 11 & 27 & 0 & 14 & 10 \\ 1 & M & 15 & 0 & 29 & 24 \\ 15 & 13 & M & 35 & 5 & 0 \\ 0 & 0 & 9 & M & 2 & 2 \\ 2 & 41 & 22 & 23 & M & 0 \\ 13 & 0 & 0 & 4 & 0 & M \end{bmatrix}$$

The branching process

At each stage of the algorithm involving the set S (the current vertex of the arborescence) of Hamiltonian cycles, an arc (i, j) has to be selected. There are two options regarding this selected arc. If the arc (i, j) is not to be included in any further discussion, then the branching is from S to the vertex $\overline{[i, j]}$ in which no Hamiltonian cycle contains the arc (i, j). If the arc (i, j) is selected, the branching is from S to $[i, j]$ in which every Hamiltonian cycle contains the selected arc. Then the lower bounds by evaluation are computed for these two vertices and the vertex with smaller lower bound is chosen for iteration. The algorithm terminates at a vertex for which the lower bound is less than or equal to the lower bounds of all the terminal vertices of the arborescence.

Selecting an arc for inclusion or exclusion

An arc (i, j) which corresponds to a zero entry is a desirable arc for inclusion. So if it is not to be considered further, there should be a penalty. For an arc (i, j) such that the (i, j)th entry in the current reduced matrix is 0, one way of defining the penalty p_{ij} is as follows. For each zero element d_{ij} in the current reduced matrix, let α_i be the smallest number in the ith row other than the identified zero d_{ij}. Likewise let β_j be the smallest number in the jth column other than the identified zero d_{ij}. If the arc (i, j) is not chosen, there should be some other arc directed from the vertex i for inclusion which implies that the penalty should be at least α_i. Similarly, there should be some arc directed to the vertex j and so the penalty should be at least β_j. Thus the penalty p_{ij} for not choosing (i, j) is $\alpha_i + \beta_j$. The number p_{ij} could be 0.

In our example, there are nine zeros in the reduced matrix. The penalties for each of these nine arcs are as follows:

- For (a, d): $\alpha_a = 10$, $\beta_d = 0$ and $p_{ad} = 10 + 0 = 10$.
- For (b, d): $\alpha_b = 1$, $\beta_d = 0$ and $p_{bd} = 1 + 0 = 1$.
- For (c, f): $\alpha_c = 5$, $\beta_f = 0$ and $p_{cf} = 5 + 0 = 5$.
- For (d, a): $\alpha_d = 0$, $\beta_a = 1$ and $p_{da} = 0 + 1 = 1$.
- For (d, b): $\alpha_d = 0$, $\beta_b = 0$ and $p_{db} = 0 + 0 = 0$.
- For (e, f): $\alpha_e = 2$, $\beta_f = 0$ and $p_{ef} = 2 + 0 = 2$.
- For (f, b): $\alpha_f = 0$, $\beta_b = 0$ and $p_{fb} = 0 + 0 = 0$.
- For (f, c): $\alpha_f = 0$, $\beta_c = 9$ and $p_{fc} = 0 + 9 = 9$.
- For (f, e): $\alpha_f = 0$, $\beta_e = 2$ and $p_{fe} = 0 + 2 = 2$.

It makes sense to choose that arc (i, j) for which the penalty is a maximum, and then the subsequent branching process involves the rejection of this arc or its inclusion. This maximum value of the penalty is denoted by t_{ij}.

In our example, the arc with the largest penalty is the arc (a, d) and the associated penalty t_{ij} is 10.

Suppose S is the current vertex of the arborescence. S is a set of Hamiltonian cycles in G. Initially $S = \Omega$. The evaluation by lower bound for S is $\delta(S)$. Let (i, j) be the arc with the maximum penalty t_{ij} associated with the current reduced matrix. If the arc (i, j) is not selected for inclusion, the branching is from S to the vertex $[\overline{i, j}]$ which is the set of all Hamiltonian cycles in S that do not contain the arc (i, j). The evaluation by lower bound for this vertex is $\delta(S) + t_{ij}$.

In our example, if the arc (a, d) with the maximum penalty is not selected, the evaluation by lower bound at the vertex $[\overline{a, d}]$ of the arborescence is $48 + 10 = 58$. The arborescence for this example is displayed in Fig. 3.2.

On the other hand, if the arc (i, j) with maximum penalty is chosen we can delete row i and column j from the current reduced $k \times k$ matrix (corresponding to S) resulting in a $(k - 1) \times (k - 1)$ matrix. The newly selected arc (i, j) is part of a connected path from p to q consisting of arcs already chosen. It is possible that $p = i$ and $q = j$. To avoid a subtour, the arc (q, p) should be eliminated. So the (q, p)th entry in the $(k - 1) \times (k - 1)$ matrix is assigned the value M. Suppose h' is the sum of the nonnegative numbers needed in reducing this matrix. Then $\delta([i, j]) = \delta(S) + h'$.

In our example, if the arc (a, d) is selected for inclusion, we delete row a and column d from the current reduced matrix $\mathbf{D}$ to obtain a 5×5 matrix denoted by $\mathbf{A}(a, d)$. Since (a, d) is chosen, the arc (d, a) has to be excluded to avoid any subtours at this stage. So the (d, a) entry is M.

$$
\mathbf{A}[a, d] = \begin{array}{c} \\ b \\ c \\ d \\ e \\ f \end{array}
\begin{array}{c} a \quad b \quad c \quad e \quad f \end{array}
\left[\begin{array}{ccccc}
1 & M & 15 & 29 & 24 \\
15 & 13 & M & 5 & 0 \\
M & 0 & 9 & 2 & 0 \\
2 & 41 & 22 & M & 0 \\
13 & 0 & 0 & 0 & M
\end{array} \right]
$$

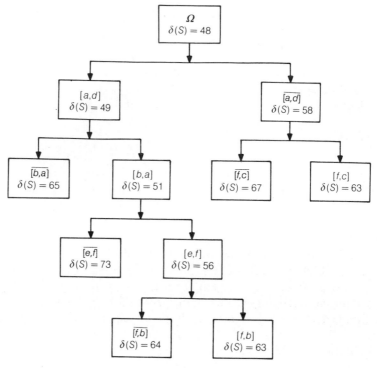

Fig. 3.2 A branch-and-bound arborescence leading to an optimal Hamiltonian cycle by reducing the cost matrix.

To evaluate the lower bound for this matrix, the matrix $\mathbf{A}([a,d])$ is reduced as before to obtain the current reduced matrix $\mathbf{D}([a,d])$ as shown below:

$$
\mathbf{D}([a,d]) = \begin{array}{c} \\ b \\ c \\ d \\ e \\ f \end{array} \begin{array}{c} \begin{matrix} a & b & c & e & f \end{matrix} \\ \begin{bmatrix} 0 & M & 14 & 28 & 23 \\ 15 & 13 & M & 5 & 0 \\ M & 0 & 9 & 2 & 2 \\ 2 & 41 & 22 & M & 0 \\ 13 & 0 & 0 & 0 & M \end{bmatrix} \end{array}
$$

In this case $h = 1$ and thus $\delta([a,d]) = 48 + 1 = 49$. Since this evaluation is less than the evaluation at the other vertex, the iteration starts from the vertex $[a,d]$ which now becomes the current vertex S.

Let us continue the iteration, starting from the vertex $S = [a,d]$ with $\delta(S) = 49$. There are seven 0s in the current reduced matrix. The penalties for these arcs are as follows:

- For (b, a): $\alpha_b = 14$, $\beta_a = 2$ and $p_{ba} = 16$.
- For (c, f): $\alpha_c = 5$, $\beta_f = 0$ and $p_{cf} = 5$.
- For (d, b): $\alpha_d = 2$, $\beta_b = 0$ and $p_{db} = 2$.
- For (e, f): $\alpha_e = 2$, $\beta_f = 0$ and $p_{ef} = 2$.
- For (f, b): $\alpha_f = 0$, $\beta_b = 0$ and $p_{fb} = 0$.
- For (f, c): $\alpha_f = 0$, $\beta_c = 9$ and $p_{fc} = 9$.
- For (f, e): $\alpha_f = 0$, $\beta_e = 5$ and $p_f = 5$.

Thus the arc (b, a) is under consideration with penalty 16. If (b, a) is not included, then the value of the lower bound is $49 + 16 = 65$. If (b, a) is included, we delete row b and column a from the current 5×5 matrix to get the following 4×4 matrix:

$$
\begin{array}{c}
 \\ c \\ d \\ e \\ f
\end{array}
\begin{array}{c}
b \quad\ c \quad\ e \quad\ f \\
\left[\begin{array}{cccc}
13 & M & 5 & 0 \\
M & 9 & 2 & 2 \\
41 & 22 & M & 0 \\
0 & 0 & 0 & M
\end{array} \right]
\end{array}
$$

The matrix given above is reduced by subtracting 2 from row d. The reduced matrix is as follows:

$$
\begin{array}{c}
 \\ c \\ d \\ e \\ f
\end{array}
\begin{array}{c}
b \quad\ c \quad\ e \quad\ f \\
\left[\begin{array}{cccc}
13 & M & 5 & 0 \\
M & 7 & 0 & 0 \\
41 & 22 & M & 0 \\
0 & 0 & 0 & M
\end{array} \right]
\end{array}
$$

Thus the lower bound at this stage is $49 + 2 = 51$. Since 51 is less than 65, $[b, a]$ becomes our current vertex S.

By inspection of this matrix, it is seen that the largest penalty is 22 when (e, f) is not considered for inclusion, giving a lower bound by evaluation of $51 + 22 = 73$.

If (e, f) is considered for inclusion, row e and column f are deleted from the reduced 4×4 matrix given above to obtain a 3×3 matrix. We are now including (e, f) along with (a, d) and (b, a). So in order to avoid subtours, we should delete the arc (f, e). The 3×3 matrix is

$$
\begin{array}{c}
 \\ c \\ d \\ f
\end{array}
\begin{array}{c}
b \quad\ c \quad\ e \\
\left[\begin{array}{ccc}
13 & M & 5 \\
M & 7 & 0 \\
0 & 0 & M
\end{array} \right]
\end{array}
$$

To reduce this matrix, we subtract 5 from row c, to give:

$$
\begin{array}{c}
 \\ c \\ d \\ f
\end{array}
\begin{array}{c}
b \quad\ c \quad\ e \\
\left[\begin{array}{ccc}
8 & M & 0 \\
M & 7 & 0 \\
0 & 0 & M
\end{array} \right]
\end{array}
$$

So the current lower bound is $51 + 5 = 56$. Since 56 is less than 73, we branch out from the vertex S with $\delta(S) = 56$.

The maximum penalty is 8 for arcs (f, b) and (c, e). The tie is broken arbitrarily. Suppose the branching is based on the arc (f, b).

If (f, b) is not included, the lower bound is $56 + 8 = 64$. If (f, b) is included, row f and row b are deleted from the reduced matrix given above. The arc (f, b) is adjoined with the set consisting of (e, f), (b, a) and (a, d). To avoid a subtour the (d, e) element becomes M.

At this stage we have the following 2×2 matrix:

$$
\begin{array}{c}
\\ c \\ d
\end{array}
\begin{array}{cc}
c & e \\
\left[\begin{array}{cc}
M & 0 \\
7 & M
\end{array}\right]
\end{array}
$$

This matrix is reduced by subtracting 7 from column c, to give:

$$
\begin{array}{c}
\\ c \\ d
\end{array}
\begin{array}{cc}
c & e \\
\left[\begin{array}{cc}
M & 0 \\
0 & M
\end{array}\right]
\end{array}
$$

So the current lower bound by evaluation is $56 + 7 = 63$. From the last 2×2 matrix, we obtain two more arcs – (c, e) and (d, c) – to complete the Hamiltonian cycle $c \to e \to f \to b \to a \to d \to c$ with a total weight of $5 + 5 + 5 + 7 + 16 + 25 = 63$.

(Suppose instead of including the arc (f, b), while breaking the tie, it was decided to include the arc (c, e). Then the resulting 2×2 matrix obtained by deleting row c and column e is

$$
\begin{array}{c}
\\ d \\ f
\end{array}
\begin{array}{cc}
b & c \\
\left[\begin{array}{cc}
M & 0 \\
0 & 0
\end{array}\right]
\end{array}
$$

The remaining two arcs to be chosen are (d, c) and (f, b), giving the optimal Hamiltonian cycle as before.)

The lower bound at the current vertex is 63, less than the lower bound at all vertices except $\lceil a, d \rceil$. So we branch out from this vertex which is our current vertex S with $\delta(S) = 58$. Now if (a, d) is not to be included, we have to select another arc from the remaining set of arcs with heavy penalty. The choice falls on (f, c), with penalty 9. So if (f, c) also is excluded, the lower bound is $58 + 9 = 67$. If (f, c) is included, we delete row f and column c from the current reduced matrix and make the (c, f) element equal to M. The resulting 5×5 matrix is as follows:

$$
\left[\begin{array}{ccccc}
M & 11 & 0 & 14 & 10 \\
1 & M & 0 & 29 & 24 \\
15 & 13 & 35 & 5 & M \\
0 & 0 & M & 2 & 2 \\
2 & 41 & 23 & M & 0
\end{array}\right]
$$

To reduce this matrix, we have to subtract 5 from row c. Thus the lower bound is $58+5$, which does not exceed the lower bound already attained. Thus the weight of an optimal Hamiltonian cycle for this six-city problem is 63.

3.3 MEDIANS AND CENTERS

Many social and industrial organizations are often faced with the problem of locating public facilities such as hospitals, schools and fire stations so as to serve the needs of the community as economically and as efficiently as possible. A **facility location problem** is a problem in which a decision has to be made regarding the exact location of a facility in the community with this optimality criterion in mind. Many such location problems can be modeled as network optimization problems where the location of a facility could be at one or more vertices of the network.

An important criterion in determining the optimal location of a facility is the way that facility is to be used once it is located. For example, if we are planning to locate a post office in a rural county consisting of a few townships as delivery areas it is desirable to locate it so that the sum of the distances from the post office to all the townships is as small as possible. A location problem of this nature, in which the aim is to minimize the sum of the travel distances, is known as a **minsum problem**. On the other hand, suppose we are interested in locating a fire station in the county. Now it makes sense to locate it so that the distance from the fire station to the farthest township is minimized. This type of problem, where the aim is to minimize the maximum travel distance, is known as a **minmax problem**.

Minsum (median) problems

Let $G = (V, E)$ be a network in which a nonnegative weight $a(i, j)$ is associated with each arc (i, j) or each edge $\{i, j\}$, as the case may be. Assume that $V = \{1, 2, \ldots, n\}$, and that there are m arcs (edges) in the network. The shortest distance in G from vertex i to vertex j is denoted by $d(i, j)$, and $\mathbf{D} = [d(i, j)]$ is the $n \times n$ shortest distance matrix. The row sum of the ith row in $\mathbf{D}$ is denoted by $s(i)$, which is the sum of the shortest distances from i to every other vertex. The number $s(i)$ is known as the **status** of the vertex i. A vertex j is called a **median vertex** if $s(j) \leqslant s(i)$ for every vertex i. In other words, a median vertex is a vertex with the smallest possible total distance from that vertex to all other vertices. For example, for a central warehouse facility from which provisions routinely have to be transported to a few other locations in the community, it is desirable to locate the facility at the median vertex in the network defined by these locations. The **median** of the network is the set of all median vertices.

In a more general setting, we can associate a nonnegative weight $w(i)$ to each vertex of the network in addition to the weight function associated with each arc. In that case the **weighted status** of the vertex i is $s(i) = w(1)d(i, 1) +$

$w(2)s(i, 2) + \cdots + w(n)d(i, n)$ and a vertex j is a **weighted median vertex** if $s(j) \leqslant s(i)$ for every vertex j. If we multiply the ith column of the SD matrix by $w(i)$ for each i, the weighted status $s(j)$ is the row sum of the modified matrix. In this case, the **weighted median** is the set of all weighted median vertices.

In some problems, it may happen that even though a facility is to be located at a vertex, the clientele who use the facility need not be at a vertex. In particular, it may happen that the clientele could be at any point in between two vertices that are joined by an arc or an edge. In this case we define the SD from a vertex to an edge or arc as follows. Suppose e is an arc from p to q. Then the SD from i to e is defined as $d(i, p) + a(p, q)$ and is denoted by $d'(i, e)$. If $e = \{p, q\}$ is an edge (undirected), then the SD from i to e is $\frac{1}{2}\{d(i, p) + d(i, q) + a(p, q)\}$. We now have the $n \times m$ matrix $\mathbf{D}'$ giving the vertex–arc shortest distance. The row sum $s(i)$ in $\mathbf{D}'$ of the ith row is the **general status** of the vertex i. A vertex j is a **general median vertex** if $s(j) \leqslant s(i)$ for every vertex i and the **general median** is the set of all general median vertices in the network.

Once the SD matrix is computed it is a trivial matter to find the median, weighted median and general median in a network.

Minmax (center) problems

Let $\mathbf{D}$ be the $n \times n$ shortest distance matrix in a network. The set of vertices is $\{1, 2, \ldots, n\}$. For each vertex i, define the **eccentricity** of i, denoted by $e(i)$, as the largest value in the ith row of the matrix $\mathbf{D}$. A vertex j is called a **center vertex** if $e(j) \leqslant e(i)$ for every vertex i in the network. In other words, a center vertex is a vertex whose farthest vertex is as close as possible. If the facility is a fire station, the best location for the facility is obviously at a center vertex. The set of all center vertices is called the **center** of the network. As in the case of medians, there are analogous definitions of a weighted center and a generalized center.

Here is a simple example to show that the center and the median of a graph can be disjoint.

Example 3.5
Find the center and median of the tree $G = (V, E)$, where $V = \{1, 2, 3, 4, 5, 6, 7\}$ and $E = \{(1, 2), (2, 3), (3, 4), (4, 5), (4, 6)$ and $(4, 7)\}$. Assume the weight of each edge is 1.

The status of the seven vertices is: $s(1) = 18$; $s(2) = 13$; $s(3) = 10$; $s(4) = 9$; and $s(5) = s(6) = s(7) = 14$. Thus the only median vertex is vertex 4.

The eccentricities of the seven vertices are: $e(1) = 6$; $e(2) = 5$; $e(3) = 2$; $e(4) = 3$; $e(5) = e(6) = e(7) = 4$. The only center vertex is vertex 3.

3.4 THE STEINER TREE PROBLEM

The Steiner tree problem is a variant of the minimal spanning tree problem. Given a subset W of vertices in a graph G with a positive weight function defined on its edges, the problem is to find a subgraph $H = (W', F)$ such that H is a tree; the weight of H is minimum; and W is a subset of W'. The subgraph satisfying these three properties is a **Steiner tree** in G with respect to the set W. A Steiner tree with respect to a pair of vertices is obviously a shortest path between them. At the other extreme, a Steiner tree with respect to V in the graph $G = (V, E)$ is a minimal spanning tree in G. If W is a proper subset of V, it is easy to see that in general a minimal spanning tree on the subgraph of G induced by W need not be a Steiner tree for the set W. Thus the efficient algorithms available to obtain a minimal spanning tree in a connected network are not directly applicable to the problem of finding a Steiner tree with respect to an arbitrary set of vertices. Unlike the optimal spanning tree problem, there is no efficient algorithm for solving this problem.

The Steiner tree problem arises in many practical optimization problems such as the design of pipeline networks or rural road networks. The vertices in the tree which are in W are called the **customer vertices** or **compulsory vertices** of T, whereas the vertices in $S = W' - W$ are called the **Steiner points** or **optional vertices**. The set S is a set of Steiner points corresponding to the set W.

Spanning tree enumeration algorithm

In this section we first discuss a method for obtaining a Steiner tree with respect to a given set W by a 'restricted' enumeration method. We examine certain subsets of vertices which contain W as a subset, obtain minimal spanning trees on graphs induced by these subsets and then choose a tree of minimum weight which will be a Steiner tree for the set W.

A complete graph with a positive weight function $a(i, j)$ for each edge (i, j) is said to have the triangle property if the triangle inequality $a(u, v) \leqslant a(u, w) + a(v, w)$ is satisfied for all vertices u, v and w in the graph.

Theorem 3.3
If the number of customer vertices in a complete graph with the triangle property is m, then the number of Steiner points cannot exceed $m - 2$.

Proof
Let the number of Steiner points be p. Then the number of vertices in the Steiner tree is $m + p$. So the number of edges in the Steiner tree is $m + p - 1$.

Let the average degree of a Steiner point be x, and let the average degree of a customer vertex be y. Then

$$px + my = 2(m + p - 1)$$

Now the triangle property implies that $x \geqslant 3$. Also, y is at least 1. So

$$2m + 2p - 2 \geqslant 3p + m$$

implying $p \leqslant m - 2$.

An arbitrary network need not be complete. Even if it is complete it need not have the triangle property. So given an arbitrary connected network $G = (V, E)$ with n vertices, we first construct the network $G' = (V, E')$ in which the weight of the edge (i, j) is the shortest distance between i and j. Then G' is a complete network with the triangle property.

In view of Theorem 3.3, to obtain a Steiner tree (in a complete graph G with n vertices with the triangle property) with respect to a set W with m vertices, it is enough if we consider those sets of the form $W \cup S$, where S is a set of vertices such that $W \cap S$ is empty and $1 \leqslant |S| \leqslant m - 2$. The number of such sets will be $1 + {}^{n-m}C_1 + {}^{n-m}C_2 + \cdots + {}^{n-m}C_{m-2}$, where ${}^{n}C_r$ is the number of ways of choosing r objects from a set of n objects. This procedure for finding a Steiner tree, due to E. Lawler, is known as the **spanning tree enumeration algorithm** and is obviously not very practical when the number of vertices is large.

Here is a simple example to illustrate this procedure in a network with seven vertices with (a) four customer vertices and (b) three customer vertices.

Example 3.6
Find the weight of the Steiner tree in the network $G = (V, E)$, shown below, whose weight matrix is given by

$$
A = \begin{bmatrix}
0 & 4 & 8 & 4 & 1 & 2 & M \\
4 & 0 & 2 & M & 2 & M & 3 \\
8 & 2 & 0 & 3 & M & M & 6 \\
4 & M & 3 & 0 & M & 7 & M \\
1 & 2 & M & M & 0 & 5 & 1 \\
2 & M & M & 7 & 5 & 0 & 5 \\
M & 3 & 6 & M & 1 & 5 & 0
\end{bmatrix}
$$

and the set W of customer vertices is (a) $W = \{1, 2, 3, 4\}$ and (b) $W = \{3, 6, 7\}$.

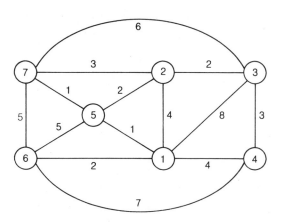

The network $G' = (V, E')$ has the following weight matrix which is the SD matrix of G:

$$D = \begin{bmatrix} 0 & 3 & 5 & 4 & 1 & 2 & 2 \\ 3 & 0 & 2 & 5 & 2 & 5 & 3 \\ 5 & 2 & 0 & 3 & 4 & 7 & 5 \\ 4 & 5 & 3 & 0 & 5 & 6 & 6 \\ 1 & 2 & 4 & 5 & 0 & 3 & 1 \\ 2 & 5 & 7 & 6 & 3 & 0 & 4 \\ 2 & 3 & 5 & 6 & 1 & 4 & 0 \end{bmatrix}$$

(a) The set of customer vertices is $W = \{1, 2, 3, 4\}$. Here $m = 4$, $m - 2 = 2$, $n - m = 3$ and $V - W = \{5, 6, 7\}$.

(i) The weight of a minimal spanning tree induced by the set $\{1, 2, 3, 4\}$ on G' is 8.

(ii) The weight of a minimal spanning tree induced by the set $\{1, 2, 3, 4, 5\}$ on G' is 8.

(iii) The weight of a minimal spanning tree induced by the set $\{1, 2, 3, 4, 6\}$ on G' is 10.

(iv) The weight of a minimal spanning tree induced by the set $\{1, 2, 3, 4, 7\}$ on G' is 10.

(v) The weight of a minimal spanning tree induced by the set $\{1, 2, 3, 4, 5, 6\}$ on G' is 9.

(vi) The weight of a minimal spanning tree induced by the set $\{1, 2, 3, 4, 5, 7\}$ on G' is 8.

(vii) The weight of a minimal spanning tree induced by the set $\{1, 2, 3, 4, 6, 7\}$ on G' is 11.

Thus the weight of a Steiner tree with W as its set of customer vertices is 8. The edges in G' of the Steiner tree are $\{1, 2\}$, $\{2, 3\}$ and $\{3, 4\}$, with weights 3, 2 and 3, respectively. The edge $\{1, 2\}$ in G' corresponds to the path 1—5—2 in G. The other two edges remain unaffected. Thus a Steiner tree with customer vertices $1, 2, 3$ and 4 has one Steiner point at vertex 5, as shown below: the Steiner point at vertex 5 is shown with two concentric circles.

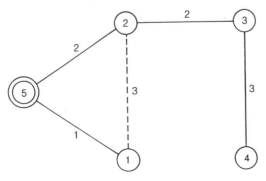

(b) The set of customer vertices is $\{3, 6, 7\}$. Here $m = 3$, $m - 2 = 1$, $n - m = 4$, $V - W = \{1, 2, 4, 5\}$.

(i) The weight of a minimal spanning tree induced by the set $\{3, 6, 7\}$ on G' is 9.
(ii) The weight of a minimal spanning tree induced by the set $\{3, 6, 7, 1\}$ on G' is 9.
(iii) The weight of a minimal spanning tree induced by the set $\{3, 6, 7, 2\}$ on G' is 9.
(iv) The weight of a minimal spanning tree induced by the set $\{3, 6, 7, 4\}$ on G' is 12.
(v) The weight of a minimal spanning tree induced by the set $\{3, 6, 7, 5\}$ on G' is 8.

Thus the weight of a Steiner tree with customer vertices at 3, 6 and 7 is 8. The edges in G' corresponding to the Steiner tree are $\{3, 5\}$, $\{5, 6\}$ and $\{5, 7\}$, with weights 4, 3 and 1, respectively. The path in G which corresponds to the edge $\{3, 5\}$ is 3—2—5. The path in G which corresponds to the edge $\{5, 6\}$ is 5—1—6. The edge $\{5, 7\}$ is the same in both G and G'. Thus a Steiner tree with customer vertices at 3, 6 and 7 has Steiner points at 1, 2 and 5 as shown below.

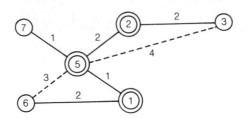

Distance network heuristic

Even though there are many exact algorithms for solving the Steiner tree problem, problems with large number of vertices require the use of approximate methods. We now discuss an approximate algorithm as outlined by Winter (1987).

Step 1. Given $G = (V, E)$ and the set W of customer vertices, construct the complete graph $H = (W, F)$ in which the weight of $\{i, j\}$ is the SD between i and j in G.
Step 2. Obtain a minimal spanning tree $T(H)$ in H.
Step 3. Replace each edge $\{i, j\}$ of $T(H)$ by a shortest path between i and j in G. Tie-breaking is arbitrary. The resulting graph, G'', is a subgraph of G.
Step 4. Obtain a minimum spanning tree $T(G'')$ in G''.
Step 5. If v is a vertex of degree 1 in $T(G'')$ and if v is not in W, delete v from the tree $T(G'')$. Continue this process by deleting one vertex at a time.

Example 3.7
Apply the distance network heuristic to the network of Example 3.6 with (a)
$W = \{1, 2, 3, 4\}$ and (b) $W = \{3, 6, 7\}$.

(a) $W = \{1, 2, 3, 4\}$. The edges in the minimal spanning tree in H are $\{1, 2\}, \{2, 3\}$
    and $\{3, 4\}$. Replace the edge $\{1, 2\}$ by the path 1—2—5. Thus a minimal
    spanning tree in G'' is 1—5—2—3—4. The degree of vertex 5 (which is
    not in W) is more than 1. Thus we have the tree 1—5—2—3—4, with 5 as
    the Steiner point. The weight of this tree is 8. So the weight of a Steiner tree
    with $W = \{1, 2, 3, 4\}$ as the set of vertices cannot exceed 8.

(b) $W = \{3, 6, 7\}$. The minimal spanning tree on H has edges $\{3, 7\}$ and $\{7, 6\}$.
    Replace these edges by their shortest paths in G to obtain a minimal
    spanning tree in G''. The edges of this tree are $\{3, 2\}, \{2, 7\}, \{7, 5\}, \{5, 1\}$ and
    $\{1, 6\}$. The vertices not in W are 1, 2 and 5. The degree of each of them is
    more than 1. The weight of this tree is 9. So the weight of a Steiner tree with
    $W = \{3, 6, 7\}$ as the set of customer vertices cannot exceed 9. An approxi-
    mate solution is the tree shown below, with weight 9. Observe that the
    approximate algorithm selected the edge $\{2, 7\}$, with weight 3, rather than
    the edge $\{2, 5\}$, with weight 2, selected by the exact algorithm, even though
    the set of Steiner points is the same in the two cases.

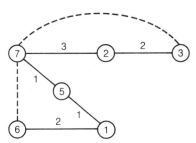

3.5 EXERCISES

1. Find the shortest distance from vertex 1 to every other vertex in the
   digraph with weight matrix **A** as given below. Draw the shortest distance
   tree rooted at vertex 1.

$$\mathbf{A} = \begin{bmatrix} 0 & 4 & 6 & 8 & - & - & - \\ - & 0 & 1 & - & 7 & - & - \\ - & - & 0 & 2 & 5 & 4 & - \\ - & - & - & 0 & - & 5 & - \\ - & - & - & - & 0 & 7 & 6 \\ - & - & - & - & 1 & 0 & 8 \\ - & - & - & - & - & - & 0 \end{bmatrix}$$

2. Find the shortest distance from vertex 1 to every other vertex in the
   digraph with weight matrix **A** as given below. Draw the shortest distance
   tree rooted at vertex 1.

$$
\mathbf{A} = \begin{bmatrix}
0 & - & 4 & 10 & 3 & - & - \\
- & 0 & 1 & 1 & 2 & 11 & 0 \\
- & 9 & 0 & 8 & 3 & 2 & 1 \\
- & 4 & 0 & 0 & 8 & 6 & 3 \\
- & 0 & 1 & 2 & 0 & 3 & 1 \\
- & 1 & 1 & 3 & 2 & 0 & 0 \\
- & 4 & 3 & - & - & 2 & 0
\end{bmatrix}
$$

3. Find the SD matrix **D** and the SP matrix **P** in the graph with weight matrix **A** as given below.

$$
\mathbf{A} = \begin{bmatrix}
0 & 1 & - & - & - & 1 & 4 \\
1 & 0 & 2 & - & - & - & 1 \\
- & 2 & 0 & 2 & - & - & 4 \\
- & - & 2 & 0 & 3 & - & - \\
- & - & - & 3 & 0 & 9 & 3 \\
1 & - & - & - & 9 & 0 & - \\
4 & 1 & 4 & - & 3 & - & 0
\end{bmatrix}
$$

4. In the matrix of Exercise 2 above, replace rows 2, 5 and 6 by $[-\ 0\ -1\ -1\ 2\ 11\ 0]$, $[-\ 0\ 1\ 2\ 0\ 3\ -1]$ and $[-\ -1\ -1\ 3\ 2\ 0\ 0]$, respectively. Find the SD matrix **D** and SP matrix **P** for the new weight matrix.

5. In Exercise 4 above, find a shortest path and the shortest distance from vertex 4 to vertex 2.

6. In Exercise 4 above construct an SD tree rooted at vertex 1.

7. In Exercise 4, find the negative cycle in the network if the number -1 which appears in the fourth column is replaced by -3.

8. In Exercise 4, find a shortest path from vertex 4 to vertex 2 which does not pass through vertices 5, 6 and 7.

9. Find a shortest path from vertex 1 to vertex 6 which does not pass through vertices 4 and 5 in the network whose weight matrix is as follows:

$$
\mathbf{A} = \begin{bmatrix}
0 & 20 & 15 & 4 & 3 & - \\
20 & 0 & 19 & - & - & 9 \\
15 & 19 & 0 & 8 & - & 10 \\
4 & - & 8 & 0 & 6 & 9 \\
3 & - & - & 6 & 0 & 7 \\
- & 9 & 10 & 9 & 7 & 0
\end{bmatrix}
$$

10. In Exercise 9, to find the SD from vertex 1 to the other vertices, take a feasible tree solution consisting of the edges (1, 2), (1, 3), (1, 4), (1, 5) and (2, 6) and perform one iteration using the network simplex method. What is the distance from 1 to 6 at this stage?

11. Suppose $G = (V, E)$ is a digraph where $V = \{1, 2, \ldots, n\}$. The weight $w(i, j)$ of the arc (i, j) need not be nonnegative. Let $d(i)$ be the shortest distance

from vertex 1 to vertex i. Show that

$$d(i) = \min\{d(j) + w(j, i) : (j, i) \in E, j \in V\}$$

(These equations are known as **Bellman's equations**.)

12. Show that a digraph is acyclic if and only if its vertices can be labeled such that $i < j$ for all arcs (i, j).

13. Show that in the case of an acyclic network, Bellman's equations (in Exercise 11 above) can be modified to

$$d(i) = \min\{d(j) + w(j, i) : j < i\}$$

14. Use Bellman's equations to find the SD from vertex 1 to the other vertices in the acyclic digraph with weight matrix

$$\begin{bmatrix}
0 & - & 3 & 1 & - & 4 & - & - \\
- & 0 & -3 & - & 1 & - & - & -4 \\
- & - & 0 & 1 & 1 & - & - & - \\
- & - & - & 0 & 6 & -1 & -1 & - \\
- & - & - & - & 0 & - & -1 & -1 \\
- & - & - & - & - & 0 & 2 & - \\
- & - & - & - & - & - & 0 & -2 \\
- & - & - & - & - & - & - & 0
\end{bmatrix}$$

15. Relabel vertex A in the acyclic digraph shown below as vertex 1. Relabel the other vertices such that whenever (i, j) is an arc $i < j$. Find the

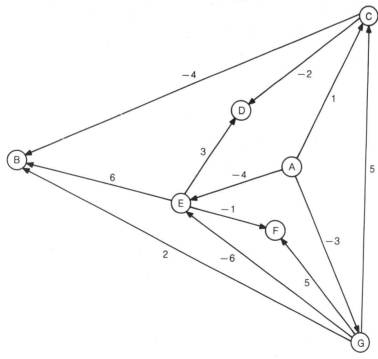

shortest distances from vertex A to the other vertices using Bellman's equations.

16. Relabel the vertices in the digraph shown below as $A = 1$ and $i < j$ whenever (i, j) is an arc. Find the SD from A to the other vertices using Bellman's equations.

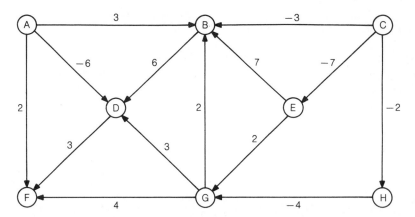

17. The 'cross product' of the $n \times n$ matrix $\mathbf{A} = [a_{ij}]$ with the $n \times n$ matrix $\mathbf{B} = [b_{ij}]$ is the $n \times n$ matrix $(\mathbf{A} \otimes \mathbf{B}) = [p_{ij}]$, where $p_{ij} = \min_k \{a_{ik} + b_{kj}\}$.
(a) Show that the cross product operation is not commutative but associative. (b) Show that if $\mathbf{U}^{(0)}$ is the $n \times n$ matrix in which each diagonal number is 0 and each nondiagonal number is $+\infty$, then $\mathbf{U}^{(0)} \otimes \mathbf{A} = \mathbf{A} \otimes \mathbf{U}^{(0)} = \mathbf{A}$.

18. Show that the cross product operation defined in Exercise 17 above can be used to obtain the SD between every pair of vertices in an arbitrary digraph with a given weight matrix.

19. Apply the cross product multiplication method to compute the SD matrix of the digraph in Exercise 4 above.

20. Obtain (i) an optimal Hamiltonian cycle and (ii) an optimal salesman circuit in $G = (V, E)$ where $V = \{1, 2, 3, 4\}$ with weight matrix

$$\begin{bmatrix} - & 5 & 19 & 11 \\ - & - & 4 & 7 \\ - & 5 & - & 14 \\ 9 & - & 6 & - \end{bmatrix}$$

by using the branch and bound method along with the assignment procedure.

21. Solve Exercise 20 above by 'reducing' the weight matrix and without using the assignment procedure.

22. Obtain (i) an optimal Hamiltonian cycle and (ii) an optimal salesman circuit in the network with weight matrix

$$\begin{bmatrix} - & 2 & - & 5 & - & - \\ - & - & 1 & - & 2 & 1 \\ 2 & - & - & - & - & 5 \\ 4 & - & - & - & - & 2 \\ - & 9 & - & 2 & - & - \\ - & - & 2 & - & 2 & - \end{bmatrix}$$

23. Obtain an optimal Hamiltonian cycle in the network with the following weight matrix:

$$\begin{bmatrix} - & 12 & 10 & 9 & 10 & 13 & 9 \\ 10 & - & 17 & 10 & 10 & 11 & 10 \\ 9 & 14 & - & 11 & 9 & 12 & 11 \\ 11 & 10 & 11 & - & 12 & 11 & 10 \\ 9 & 11 & 11 & 9 & - & 14 & 12 \\ 10 & 10 & 10 & 10 & 10 & - & 10 \\ 9 & 10 & 9 & 10 & 9 & 10 & - \end{bmatrix}$$

24. Find the median, the general median, the center and the general center of the graph of Exercise 3.
25. Find the weighted median of the graph in Exercise 3 if the weights of the vertices are $0, 0, 0, 1, 0, 0$ and 0, respectively.
26. Find a Steiner tree corresponding to the set W of customer vertices in the graph in Exercise 3 where $W = \{1, 3, 5\}$.

4

Minimum cost flow problems

4.1 THE UPPER-BOUNDED TRANSSHIPMENT PROBLEM

In the transshipment problems of Chapter 2, there was no restriction on a feasible flow vector other than the requirement that it is nonnegative. There are some combinatorial optimization problems (as in aircraft scheduling) which can be formulated as network flow problems only if the flow along each arc (i, j) is bounded below by a nonnegative lower bound v_{ij} and bounded above by an upper bound u_{ij}. So if $\mathbf{A}$ is the incidence matrix and $\mathbf{b}$ is the supply–demand vector, a vector $\mathbf{x}$ is a feasible solution if and only if $\mathbf{Ax} = \mathbf{b}$ and $\mathbf{v} \leqslant \mathbf{x} \leqslant \mathbf{u}$. If $\mathbf{c}$ is the cost vector, the optimization problem is the problem of finding a feasible vector $\mathbf{x}$ such that $\mathbf{cx}$ is as small as possible.

We first observe that a transshipment problem with a nonnegative lower bound constraint can be converted into a problem in which the lower bound vector is zero. Consider replacing the flow x_{ij} by $x'_{ij} + v_{ij}$ for each arc (i, j) so that the transformed flow vector $\mathbf{x}'$ becomes a nonnegative vector each component of which is bounded above by $u_{ij} - v_{ij}$. The objective function $\mathbf{cx}$ is transformed into $\mathbf{cx}' + \mathbf{cv}$, where $\mathbf{cv}$ is a constant. The component b_i is replaced by $b_i + r_i - s_i$, where r_i is the sum of the lower bounds of the arcs directed from the vertex i and s_i is the sum of the lower bounds of the arcs directed to the vertex i. Thus without loss of generality we can assume that the lower bound vector is zero.

A transshipment problem is said to be **upper-bounded** if the component of a nonnegative feasible flow vector along an arc (i, j) cannot exceed a specified positive amount u_{ij}, called the **capacity** of the arc. The vector $\mathbf{u} = [u_{ij}]$ is the **capacity vector** of the network. An upper-bounded transshipment problem is also known as a **capacitated transshipment problem** or a **minimum cost flow problem**. This problem is a generalization of the uncapacitated problem discussed in Chapter 2 and the algorithm described there can be suitably modified for solving problems with this additional capacity constraint.

If the component of a feasible flow vector $\mathbf{x}$ along an arc (i, j) is zero then the arc (i, j) is said to be **free** with respect to $\mathbf{x}$. The arc (i, j) is said to be **saturated** (or **full**) with respect to $\mathbf{x}$ if the component is equal to the capacity of the arc. An arc which is neither free nor saturated is **unsaturated**. A feasible flow vector $\mathbf{x}$ in a capacitated network with n vertices is called a **feasible tree solution** (FTS) if there exists a tree T with $n - 1$ arcs such that any arc not in the tree is either free or saturated.

As in Chapter 2, we start with an FTS $\mathbf{x}$ and obtain the unique dual solution $\mathbf{y}$ with n components by solving the defining equations $y_j - y_i = c_{ij}$ for each arc (i, j) of the tree and making use of the equilibrium condition which enables us to assign an arbitrary value to one of the components of $\mathbf{y}$. A free arc (i, j) not in T is **profitable** if $y_i + c_{ij} < y_j$. A saturated arc (i, j) not in T is profitable if $y_i + c_{ij} > y_j$.

We now have the following analogue of Theorem 2.2.

Theorem 4.1
A feasible tree solution with no profitable arcs is an optimal solution.

Proof
Let $\mathbf{x} = [x_{ij}]$ be an FTS with no profitable arcs and let $\mathbf{x}' = [x'_{ij}]$ be any feasible solution. We have to prove $\mathbf{cx} \leqslant \mathbf{cx}'$.

Let $\mathbf{d} = [d_{ij}]$ where $d_{ij} = y_i + c_{ij} - y_j$. In other words, $\mathbf{d} = \mathbf{c} - \mathbf{yA}$. So

$$\mathbf{cx} = \mathbf{dx} + \mathbf{yAx} = \mathbf{dx} + \mathbf{yb}$$
$$\mathbf{cx}' = \mathbf{dx}' + \mathbf{yAx}' = \mathbf{dx}' + \mathbf{yb}$$

so it is enough if we prove that $\mathbf{dx} \leqslant \mathbf{dx}'$. We consider three cases:

(a) (i, j) is an arc of the tree.
(b) (i, j) is not an arc of the tree and is free.
(c) (i, j) is not an arc of the tree and is saturated.

In case (a), $d_{ij} = 0$ and so $d_{ij}x_{ij} = d_{ij}x'_{ij}$.
In case (b), $d_{ij} \geqslant 0$, $x_{ij} = 0$ and so $d_{ij}x_{ij} \leqslant d_{ij}x'_{ij}$.
In case (c), $d_{ij} \leqslant 0$, $x_{ij} = u_{ij}$, $x'_{ij} \leqslant u_{ij}$. So $d_{ij}x'_{ij} \geqslant d_{ij}x_{ij}$.
Thus $\mathbf{dx} \leqslant \mathbf{dx}'$ in all cases. This completes the proof.

Thus if $\mathbf{x}$ is not optimal there is at least one arc which is profitable. In the second stage of the algorithm we select an arbitrary profitable arc e (called the entering arc) and adjoin this arc e to T to obtain a unique cycle $C(e)$. An arc in this cycle is forward if it has the same direction as e, and backward otherwise. If the entering arc is saturated, then subtract t units from the existing flow in that arc and in all forward arcs and add t units to the existing flow in all backward arcs. If the entering arc is free, then add t units to the existing flow in that arc and in all forward arcs and subtract t units from the existing flow in all backward arcs. Of course, when we add or subtract the nonnegative number t we have to make sure that the revised flow in each arc of $C(e)$ is still a nonnegative number which does not exceed the capacity of that arc. Now we choose t such that an arc f in $C(e)$ becomes either free or saturated. This arc f is a leaving arc. The arcs e and f need not be distinct. At this stage we have an updated flow vector $\mathbf{x}'$ associated with the tree T' obtained by deleting f from T and adjoining e to it. The flow $\mathbf{x}'$ thus obtained from the nonoptimal FTS $\mathbf{x}$ is called the **updated flow** and we then have the following theorem which is the analogue of Theorem 2.3. As in Chapter 2, it is assumed that the cost vector is nonnegative.

Theorem 4.2
The updated flow $\mathbf{x}'$ is an FTS and $\mathbf{cx} - \mathbf{cx}'$ is nonnegative.

Proof
When $\mathbf{x}$ is changed into $\mathbf{x}'$, the netflow at each vertex remains the same and so $\mathbf{x}'$ is a feasible flow. Furthermore, if the arc (i, j) is not an arc of the tree T' then the arc is either free or saturated. Thus $\mathbf{x}'$ is indeed an FTS.

Let T be the tree associated with $\mathbf{x}$, and let $\mathbf{y}$ be the corresponding dual solution corresponding to $\mathbf{x}$. Let $\mathbf{d} = \mathbf{c} - \mathbf{yA}$. Then $\mathbf{cx}' - \mathbf{cx} = \mathbf{dx}' - \mathbf{dx}$.

In computing $\mathbf{dx}$, the contribution from arcs in the tree T as well as from the free arcs (not in the tree) is zero. Thus $\mathbf{dx} = \Sigma d_{ij} u_{ij}$, where the summation is along the set of saturated arcs.

In computing $\mathbf{dx}'$, there are two cases to consider:

(a) The entering arc (p, q) is a free arc. Then $\mathbf{dx}' = d_{pq}(t) + \Sigma d_{ij} u_{ij}$, where the summation is along the set of all saturated arcs. Thus $\mathbf{dx}' - \mathbf{dx} = d_{pq}(t)$, where t is nonnegative and d_{pq} is negative.

(b) The entering arc (p, q) is a saturated arc. Then $\mathbf{dx}' = d_{pq}(u_{pq} - t) + \Sigma d_{ij} u_{ij}$, where the summation is along the set of all saturated arcs other than the arc (p, q). Then $\mathbf{dx}' - \mathbf{dx} = d_{pq}(u_{pq} - t) - d_{pq} u_{pq} = d_{pq}(-t)$, where t is nonnegative and d_{pq} is positive.

Thus in any case $\mathbf{cx} - \mathbf{cx}' = \mathbf{dx} - \mathbf{dx}' = t|d_{pq}|$, where (p, q) is the entering arc. This completes the proof.

We can now summarize this iteration process as follows:

Step 1. The current FTS is $\mathbf{x}$ with a tree T. Any arc not in T is either free or saturated.

Step 2. Obtain a (unique) vector $\mathbf{y}$ with n components, the last being zero, by solving the $n - 1$ defining equations of the type $y_v - y_u = c_{uv}$, where (u, v) is an arc in T.

Step 3. Let $d_{ij} = y_i + c_{ij} - y_j$. A free arc (i, j) is profitable if d_{ij} is negative. A saturated arc (i, j) is profitable if d_{ij} is positive. If there are no profitable arcs with respect to the current FTS $\mathbf{x}$, then $\mathbf{x}$ is optimal and we stop.

Step 4. Otherwise choose a profitable arc $e = (p, q)$ as the entering arc. Then a unique cycle C is created. Arcs in C which have the same direction as e are forward arcs, the others being backward arcs.

Step 5.
(a) If the entering arc e is free, then add t units along e and also to the flow along all forward arcs. At the same time subtract t units from the flow in all backward arcs.
(b) If the entering arc e is saturated, then subtract t units from the flow in e and also from the flow in all forward arcs. At the same time add t units to the flow in all backward arcs.
(c) t is nonnegative and the choice of t is subject to capacity constraints and to the nonnegativity constraints of the flow vector.

Step 6. Choose t such that in the cycle C there is at least one arc which becomes free or saturated. Remove one such arc from T, creating a new tree T' and an updated flow $\mathbf{x}'$.

Step 7. $\mathbf{x}'$ is an FTS and $\mathbf{cx}' \leqslant \mathbf{cx}$. Go to step 1.

Step 8. $\mathbf{cx} - \mathbf{cx}' = t|d_{pq}|$.

Example 4.1

Consider the upper-bounded network shown below, with 6 vertices and 11 arcs, with supply–demand vector $\mathbf{b}$ given by $\mathbf{b}^T = [-9\ 4\ 17\ 1\ -5\ -8]$. Any flow vector $\mathbf{x}$ is of the form $\mathbf{x}^T = [x_{12}\ x_{13}\ x_{15}\ x_{23}\ x_{42}\ x_{43}\ x_{53}$ $x_{54}\ x_{56}\ x_{62}\ x_{64}]$.

The cost vector $\mathbf{c}$ is given by $\mathbf{c} = [3\ 5\ 1\ 1\ 4\ 1\ 6\ 1\ 1\ 1\ 1]$, and the capacity vector $\mathbf{u}$ by $\mathbf{u}^T = [2\ 10\ 10\ 6\ 8\ 9\ 9\ 10\ 6\ 7\ 8]$.

Take $\mathbf{x}^T = [2\ 0\ 7\ 5\ 0\ 9\ 3\ 3\ 6\ 7\ 7]$ as initial FTS.

(This example is from Chvátal, 1980.)

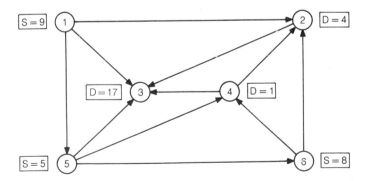

Iteration 1

Step 1. $\mathbf{x}^T = [2\ 0\ 7\ 5\ 0\ 9\ 3\ 3\ 6\ 7\ 7]$ and $z = \mathbf{cx}^T = 68$. The arcs in T are $(1, 5), (2, 3), (5, 3), (5, 4)$ and $(6, 4)$. Using these arcs we get $\mathbf{y} = [-1\ 5\ 6\ 1\ 0\ 0]$.

Step 2. The arc $(1, 3)$, which is not in T, is free and profitable because $y_1 + c_{13} = -1 + 5 = 4 < 6 = y_3$.

The arc $(5, 6)$, which is not in T, is saturated and profitable because $y_5 + c_{56} = 0 + 1 = 1 > 0 = y_6$.

The current FTS is not optimal. Either $(1, 3)$ or $(5, 6)$ can enter T. Let $e = (1, 3)$ be the entering arc.

Step 3. The cycle $C(e)$ is $1 \to 3 \leftarrow 5 \leftarrow 1$, in which $(1, 5)$ and $(5, 3)$ are backward with current flow 7 and 3, respectively. The revised flow in $C(e)$ is t units along $(1, 3)$, $3 - t$ units along $(5, 3)$ and $7 - t$ units along $(1, 5)$, where t cannot exceed the capacity of the arc $(1, 3)$. Thus $t = 3$ and the arc $f = (5, 3)$ leaves T.

Iteration 2

Step 1. $\mathbf{x}^T = [2\ 3\ 4\ 5\ 0\ 9\ 0\ 3\ 6\ 7\ 7]$ and $z = 62$. The arcs in T are $(1, 3)$, $(1, 5)$, $(2, 3)$, $(5, 4)$ and $(6, 4)$, and we have $\mathbf{y} = [-1\ 3\ 4\ 1\ 0\ 0]$.

Step 2. The arc $(5, 6)$, which is not in T, is saturated and profitable because $y_5 + c_{56} = 0 + 1 > 0 = y_6$. (This is the only profitable arc.) The current FTS is not optimal. The arc $e = (5, 6)$ is the entering arc.

Step 3. The cycle $C(e)$ is $5 \to 6 \to 4 \leftarrow 5$, in which $(6, 4)$ is forward and $(5, 4)$ is backward. The current flows in $(5, 6)$, $(6, 4)$ and $(5, 4)$ are 6, 7 and 3, respectively. The revised flow will be $6 - t$, $7 - t$ and $3 + t$, where $3 + t$ cannot exceed the capacity of $(5, 4)$, which is 10. We obtain $t = 6$, and $f = (5, 6)$ leaves the tree. (It is interesting to observe that even though $e = f$ the updated flow vector is different.)

Iteration 3

Step 1. $\mathbf{x}^T = [2\ 3\ 4\ 5\ 0\ 9\ 0\ 9\ 0\ 7\ 1]$ and $z = 56$. The arcs in T are $(1, 3)$, $(1, 5)$, $(2, 3)$, $(5, 4)$ and $(6, 4)$, and $\mathbf{y} = [-1\ 3\ 4\ 1\ 0\ 0]$.

Step 2. There are no profitable arcs. The current FTS is optimal.

4.2 INITIALIZATION AND FEASIBILITY

As in the case of uncapacitated transshipment problems, we are interested in ascertaining whether a given upper-bounded problem has a feasible solution and in obtaining an initial FTS if the problem is feasible. A feasible tree solution is readily available if we can locate a vertex v satisfying the following three properties:

1. If i is a source, then there is an arc from i to v whose capacity is not less than the supply at i.
2. If j is a sink, then there is an arc from v to j whose capacity is not less than the demand at j.
3. If k is an intermediate vertex, then there is an arc from v to k or from k to v.

A vertex satisfying these properties need not exist. Even if it exists it may not be easy to locate it. So we choose an arbitrary vertex v and construct artificial arcs if necessary to obtain an initial FTS for an enlarged network, as in the case of uncapacitated transshipment problems. The only difference between the two cases is that in the upper-bounded problem the enlarged graph may have multiple arcs. In addition to the existing arc (i, j) from i to j with a prescribed upper bound, there may be another (artificial) arc from i to j with unlimited capacity in the enlarged network.

The procedure for constructing $G' = (V, E')$ from the given network $G = (V, E)$ is quite straightforward. Choose an arbitrary vertex v. If i is a source and if there is an arc from i to v with capacity not less than the supply at i, then the arc (i, v) is in the tree. If there is no arc from i to v, then construct an artificial arc from i to v with unlimited capacity. If there is an arc from i to v with capacity less than the supply at i, then construct one more arc from i to v with unlimited capacity. Likewise if j is a sink and if there is an arc from v to j whose capacity is not less than the demand at j, then (v, j) is in the tree. If there is no arc from v to j, then construct one with unlimited capacity. If there is an arc with capacity less than the demand at j then construct one more from v to j with unlimited capacity. Finally, if k is an intermediate vertex and if there is an arc from v to k or from k to v, then that arc is chosen for the tree. Otherwise construct an artificial arc from v to k with unlimited capacity. We now have an enlarged network with $A' = E' - E$ as the set of artificial arcs for which an initial FTS $\mathbf{x}'$ can be easily obtained. Now we solve the upper-bounded transshipment problem for the enlarged network G' with cost vector $\mathbf{c}'$, the component of which along the arc (i, j) is zero if (i, j) is an original arc and one if it is an artificial arc. Let $\mathbf{x}^*$ be an optimal solution of this enlarged problem. As in Chapter 2, we have the following **infeasibility criterion**: the component of $\mathbf{x}^*$ corresponding to an artificial arc is positive if and only if the original problem is infeasible.

Example 4.2

Consider the network $G = (V, A)$ shown below, where the flow vector is $\mathbf{x}^T = [x_{12} \ x_{13} \ x_{24} \ x_{32} \ x_{34}]$, the capacity vector is $\mathbf{u}^T = [8 \ 3 \ 16 \ 1 \ 8]$, and $\mathbf{b}^T = [-7 \ 10 \ -9 \ 6]$. Obviously this is an infeasible problem.

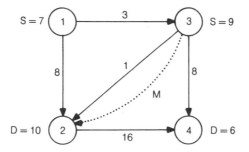

In addition to the arc $(3, 2)$ with capacity 1, an artificial arc $(3, 2)$ is constructed (drawn as a dotted line) with unlimited capacity. The enlarged flow vector is $(\mathbf{x}')^{\mathrm{T}} = [x_{12} \ x_{23} \ x_{24} \ x_{32} \ x_{34}; x_{32}]$, with cost vector $\mathbf{c}' = [0 \ 0 \ 0 \ 0 \ 0; 1]$.

Solving this problem, we obtain the optimal tree solution of the enlarged problem as $[7 \ 0 \ 0 \ 0 \ 6; 3]$, in which the flow along an artificial arc is positive.

Example 4.3

Consider the network $G = (V, A)$ shown below, where $\mathbf{x}^{\mathrm{T}} = [x_{12} \ x_{24} \ x_{31} \ x_{34}]$, $\mathbf{u}^{\mathrm{T}} = [9 \ 10 \ 2 \ 18]$ and $\mathbf{b}^{\mathrm{T}} = [-5 \ 8 \ -9 \ 6]$. This is also an infeasible problem.

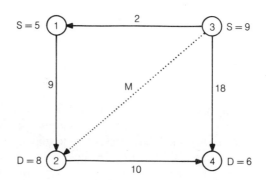

An artificial arc $(3, 2)$ is constructed with unlimited capacity and $(\mathbf{x}')^{\mathrm{T}} = [x_{12} \ x_{24} \ x_{31} \ x_{34}; x_{32}]$. The cost vector is $\mathbf{c}' = [0 \ 0 \ 0 \ 0; 1]$.

It can be easily verified that the optimal tree solution of the enlarged problem is $[5 \ 0 \ 2 \ 6; 1]$, in which the flow along an artificial arc is positive.

More on infeasibility

Let us examine infeasibility from a different perspective. Suppose W is a set of vertices in a capacitated network in which V is the set of vertices. The **total demand** at W is the sum of the demands at the vertices in W and the **total supply** at W is the sum of the supplies at the vertices in W. The **net demand** $N(W)$ at W is the total demand minus the total supply.

Let W' be the relative complement of W in V. If $\mathbf{x}$ is any feasible flow in the network, the **inflow** $I(W)$ into W is the sum of all x_{ij}, where i is in W' and j is in W. The **outflow** $O(W)$ from W is the sum of all x_{ij}, where i is in W and j is in W'. It can easily be verified (section 2.2) that $N(W)$ is equal to $I(W) - O(W)$ for any feasible flow.

The **total import capacity** $TIC(W)$ of a set W of vertices in a capacitated network $G = (V, E)$ is the sum of the capacities of all arcs (i, j) where j is in W and i is in W'. Obviously, if there is a set W of vertices such that $N(W) > TIC(W)$, then the problem is infeasible. The converse also is true and we thus have the following infeasibility criterion.

Theorem 4.3 (Gale infeasibility condition)

The capacitated transshipment problem is infeasible if and only if there is a set of vertices for which the net demand exceeds the total import capacity.

Proof

The condition is obviously sufficient.

Suppose the problem in $G = (V, A)$ is infeasible. If A' is the set of artificial arcs, then there exists an arc (p, q) in A' such that the component of the optimal tree solution x^* of the enlarged problem along that arc is positive. Here the component of the cost vector along an arc of the enlarged problem is 1 if the arc is artificial and 0 otherwise. Then $y_q - y_p = 1$. Let $W = \{k \in V: y_k \geqslant y_q\}$. Let (i, j) be any arc of the enlarged network.

Case 1: $i \in W$ and $j \in W'$. Then $y_i > y_j$. If (i, j) is an arc of the optimal tree, then $y_j - y_i$ is either 1 or 0. In any case, $y_j \geqslant y_i$. So (i, j) is not an arc of the tree. If the flow along the arc is positive, then it is a saturated arc (i.e. $x_{ij} = u_{ij}$) and it is not a profitable arc. In that case $y_i + 1 \leqslant y_j$ or $y_i \leqslant y_j$. Thus the flow along (i, j) cannot be positive. Thus the outflow $O(W)$ from W is zero for the optimal solution x^* in the enlarged network.

Case 2: $i \in W'$ and $j \in W$. Now let us consider the inflow $I(W)$ into the set W for the optimal solution x^*.

$$I(W) = \sum_{\substack{(i,j) \in E \\ i \in W', j \in W}} x_{ij}^* + \sum_{\substack{(i,j) \in A \\ i \in W', j \in W}} x_{ij}^*$$

So

$$I(W) \geqslant x_{pq}^* + \sum_{\substack{(i,j) \in E \\ i \in W', j \in W}} x_{ij}^*$$

If (i, j) is an original arc in the tree, then $y_i = y_j$. So $(i, j) \in E$, $i \in W'$ and $j \in W$ imply that (i, j) is not an arc of the tree. Thus it has to be a saturated arc. So

$$\sum_{\substack{(i,j) \subset E \\ i \in W', j \in W}} x_{ij}^* = \sum_{\substack{(i,j) \in E \\ i \in W', j \in W}} u_{ij}$$

$$= TIC(W)$$

Hence $I(W) - TIC(W) \geqslant x_{pq}^* > 0$. So $I(W) > TIC(W)$. Now $O(W) = 0$ and $N(W) = I(W) - O(W)$. Thus $N(W) > TIC(W)$. In other words, for the set W, the net demand exceeds its total import capacity.

Decomposition into subproblems

Suppose the component of x^* with respect to every artificial arc is 0. Then the original problem has a feasible solution. Let T^* be the tree associated with the optimal solution x^*. We have two possibilities:

1. No arc in T^* is artificial. So T^* is a tree in G and the subvector x induced by x^* is an FTS for the problem with T^* as the spanning tree (ignoring

directions) in G. The problem can be initialized with this FTS. See Example 2.4 for a similar situation in the case of an uncapacitated problem.

2. T^* has at least one artificial arc. Since the problem is feasible the component of $\mathbf{x}^*$ along any artificial arc in T^* is necessarily 0.

In case 2, as in Chapter 2, the problem decomposes into subproblems (see the summary at end of section 2.2). But there is a difference between the two cases. In the uncapacitated case the set V can be partitioned into two sets W and W' such that (i) the net demand of W is 0; and (ii) there are no arcs from vertices in W' to vertices in W. Then the problem can be decomposed into subproblems as in Example 2.5. In the capacitated case, as we saw in Theorem 4.3, there exists a set W such that the net demand of W is the total import capacity of W. Then V is partitioned into two sets W and W', and using this partition the problem can be decomposed into two subproblems as shown in the following example.

Example 4.4

Consider the network shown below, with flow vector $\mathbf{x}^T = [x_{12}\ x_{14}\ x_{23}\ x_{34}$ $x_{52}\ x_{56}\ x_{57}\ x_{63}\ x_{67}]$, $\mathbf{u}^T = [9\ 9\ 9\ 9\ 2\ 9\ 9\ 1\ 9]$, $\mathbf{b}^T = [-7\ -6\ 7\ 9\ -9$ $2\ 4]$ and a nonnegative cost vector.

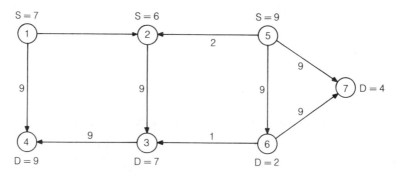

If we solve this problem by constructing artificial arcs, then it is possible to obtain the optimal tree solution shown below, where $(5, 3)$ is an artificial arc

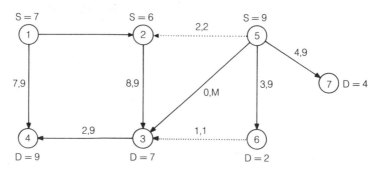

with zero flow. The arcs $(5, 2)$ and $(6, 3)$ are not in the tree but they are saturated and they are drawn as dotted lines.

Here $\mathbf{y} = [1\ 1\ 1\ 1\ 0\ 0\ 0]$ and $y_3 = 1$. $W = \{i : y_i \geq 1\} = \{1, 2, 3, 4\}$, and $N(W) = TIC(W)$. Thus the problem decomposes into two subproblems as shown below. Notice that the supply at vertex 2 is increased by 2 units (the inflow at 2) whereas the demand at vertex 3 is decreased (the inflow at 3). The optimal cost of the given problem is the sum of the optimal costs of the two subproblems plus the decomposition cost $c_{52} u_{52} + c_{63} u_{63}$.

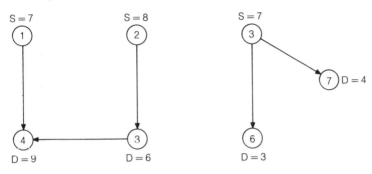

Integrality of an optimal flow

We conclude this section by observing, as in the case of the uncapacitated problem, that if the upper-bounded problem has a feasible solution then it has an optimal solution vector whose components are all integers if all the components of the supply–demand vector and all the finite components of the capacity vector are integers.

4.3 THE MAXIMUM FLOW PROBLEM

Consider an upper-bounded directed network $G = (V, E)$ where $V = \{1, 2, \ldots, n\}$. We take vertex 1 as the unique supply vertex (known as the **source**) with supply equal to v, and vertex n as the unique demand vertex (known as the **sink** or **terminal**) with demand equal to v. All the other vertices are intermediate vertices. The supply–demand vector is the $n \times 1$ vector $\mathbf{b}$ given by $\mathbf{b}^{\mathsf{T}} = [-v\ 0\ 0 \cdots 0\ v]$. Suppose G has m arcs and $\mathbf{u}$ is the capacity vector. The vertex–arc incidence matrix is the $n \times m$ matrix $\mathbf{A}$. A nonnegative vector $\mathbf{x}$ with m components is a **feasible flow** if $\mathbf{Ax} = \mathbf{b}$ and $\mathbf{x} \leq \mathbf{u}$. The problem of finding a feasible flow such that v is as large as possible is known as the maximum flow problem (MFP). The nonnegative number v is called the **value** of the flow $\mathbf{x}$. In an MFP we may assume without loss of generality the indegree of the source and the outdegree of the sink are both 0.

Since the value v of a feasible flow is a quantity not known beforehand, it is considered as a (nonnegative) variable. If p is the sum of the capacities of all the arcs directed from the source and if q is the sum of the capacities of all the arcs directed to the sink, then an upper bound of any feasible flow value is the minimum of p and q. If $v = 0$, then $\mathbf{x} = \mathbf{0}$ is a feasible flow.

In trying to maximize the flow in a given network in an arbitrary fashion, it is possible to have a nonoptimal flow pattern in the network in which an obvious method of increasing the flow may not be easily discernible. Consider, for example, the flow shown below, in which the flow value is 11.

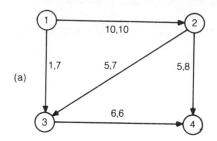

(a)

It looks as though no more flow can be sent from the source (vertex 1) to the sink (vertex 4) since the arc $(3, 4)$ is saturated. But it is certainly possible to increase the flow value in the network as can be seen below. So it is important to have a systematic procedure for solving the MFP.

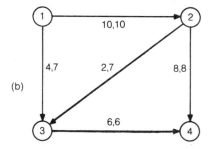

(b)

We first show that the MFP can be formulated as an upper-bounded transshipment problem as follows. Construct an arc from n to 1 (with unlimited capacity) and call it e. Let E' be the set obtained by adjoining e to E, and let $G' = (V, E')$ be the enlarged network with $m + 1$ arcs. The capacity vector of G' is $\mathbf{u}'$, of which the last component is infinite. The subvector consisting of the first m components is the capacity vector $\mathbf{u}$.

The $n \times (m + 1)$ vertex–arc incidence matrix of G' is $\mathbf{A}'$, in which the last column corresponds to the arc e. A feasible flow in G' is a nonnegative vector $\mathbf{x}'$ such that $\mathbf{A}'\mathbf{x}' = \mathbf{0}$ and $\mathbf{x}' \leqslant \mathbf{u}'$. A feasible flow in G' is also known as a **circulation** in G'. Obviously, if $\mathbf{x}'$ is a circulation in G' then the subvector $\mathbf{x}$ consisting of the first m components of $\mathbf{x}'$ is a feasible flow for the MFP for which the flow value v is the last component of $\mathbf{x}'$. The problem of maximizing v is equivalent to the problem of minimizing the negative of this last component, and this concept enables us to define a cost vector $\mathbf{c}'$ in G' as a vector in which the first m components are 0 and the last component is -1.

Thus to solve the MFP we first solve the following upper-bounded transshipment problem: find a nonnegative $\mathbf{x}'$ such that $\mathbf{c}'\mathbf{x}'$ is a minimum, with $\mathbf{A}'\mathbf{x}' = \mathbf{0}$ and $\mathbf{x}' \leqslant \mathbf{u}'$. Then the subvector $\mathbf{x}$, consisting of the first m components of $\mathbf{x}'$, is a maximum flow in G, and the last component of $\mathbf{x}'$ is the maximum flow value in G.

Notice that this is the first time that we have introduced a cost vector with negative components. The presence of negative components in the cost vector may make the problem unbounded in the sense that the optimum cost for a feasible flow could be $-\infty$. In an MFP, if we assume that each arc emanating from the source has a finite capacity then any flow value is necessarily finite and therefore the problem cannot be unbounded. Even this condition can be relaxed, as we will see later in this chapter.

The max-flow min-cut theorem

Consider a directed path P from the source to the sink in a network. Let v be value of the flow along this path. Obviously v cannot exceed the capacity of any of the arcs in P, and the maximum value of v is equal to the minimum of the capacities of the arcs in the path. If an arc in P is deleted, then we cannot send flow from source to sink using this path. In other words, each arc is a 'cut' and the capacity of an arc can be called a cut capacity. We arrive at a simple conclusion: if an upper-bounded network is a directed path then the maximum flow value in it is equal to the capacity of the minimum cut. The fundamental theorem in network flow theory, known as the **max-flow min-cut theorem**, asserts that this property is true in a more general setting.

Let $G = (V, E)$ be an upper-bounded directed network with capacity vector $\mathbf{u}$, and let $V = \{1, 2, \ldots, n\}$ be partitioned into two subsets S and T such that 1 (the source) is in S and n (the sink) is in T. Let (S, T) be the set of arcs (i, j) in E such that $i \in S$ and $j \in T$. If we delete all the arcs in (S, T) from G then we cannot send any flow from 1 to n in the network. Notice that such a deletion need not disconnect the underlying graph of the network. Thus (S, T) is called a **source-sink cut** of the network. Once the context is clear, a source-sink cut is usually just called a **cut**. An arc (i, j) in the cut (S, T) is a **forward arc of the cut** if $i \in S$ and $j \in T$ and it is a **backward arc** if $i \in T$ and $j \in S$. The sum of the capacities of all the forward arcs in the cut (S, T) is its **cut value**, denoted by $C(S, T)$. A cut (S, T) such that $C(S, T) \leqslant C(S', T')$ for any cut (S', T') in the network is called a **minimum cut** in the network and its cut value is the **minimum cut value** of the network.

Theorem 4.4

In a capacitated network the maximum flow value is equal to the minimum cut value.

Proof

The proof is in two parts: (a) no flow value in the network can exceed a cut value in the network; and (b) there exists a cut (S, T) and a flow value v such that $v = C(S, T)$.

Let $x = [x_{ij}]$ be any feasible flow vector with flow value v, and let (S, T) be any cut in the network. Then

$$
v = \sum_{i \in S} \left(\sum_j x_{ij} - \sum_j x_{ji} \right)
$$

$$
= \sum_{i \in S} \sum_{j \in S} (x_{ij} - x_{ji}) + \sum_{i \in S} \sum_{j \in T} (x_{ij} - x_{ji})
$$

$$
= \sum_{i \in S} \sum_{j \in T} (x_{ij} - x_{ji})
$$

Now $0 \leqslant x_{ij} \leqslant u_{ij}$. So $v \leqslant C(S, T)$. This completes the proof of (a). We can now conclude that if v is the maximum flow value and (S, T) is a minimum cut then v cannot exceed $C(S, T)$. It is also obvious that if v is an arbitrary flow value and (S, T) is an arbitrary cut such that v equals $C(S, T)$, then v is a maximum flow value and (S, T) is a minimum cut.

It remains to be proved, as claimed in (b), that if v is the maximum flow value then there exists a cut such that v equals its capacity. Let $\mathbf{x}$ be a feasible flow vector in the network with a finite flow value v which is a maximum. Then we prove that V can be partitioned into two sets S and T with $1 \in S$ and $n \in T$, creating a cut (S, T) such that

1. if (i, j) is a forward arc in the cut, the flow in that arc equals its capacity;
2. if (i, j) is a backward arc in the cut with $i \in T$, then the flow in that arc is 0;
3. more importantly, $v = C(S, T)$.

Construct the arc $e = (n, 1)$ with infinite capacity, and let E' be the set obtained by adjoining e to E. We thus have an enlarged upper-bounded network $G' = (V, E')$. Assume that E has m arcs. Let $\mathbf{A}'$ be the $n \times (m + 1)$ vertex–arc incidence matrix of G' in which the last column corresponds to the arc e. Let $\mathbf{u}'$ be the vector with $m + 1$ components such that the subvector $\mathbf{u}$, comprising the first m components of $\mathbf{u}'$, is the capacity vector of G and the last component is an arbitrary large positive number. A nonnegative vector $\mathbf{x}'$ such that $\mathbf{A}'\mathbf{x}' = \mathbf{0}$ and $\mathbf{x}' \leqslant \mathbf{u}'$ is called a circulation in G'. The problem of finding a circulation is always feasible because $\mathbf{x}' = \mathbf{0}$ is a circulation. Obviously, the last component of any circulation in G' is a flow value in G. Let $\mathbf{c}'$ be a row vector with $m + 1$ components in which the first m components are 0 and the last one is -1. A circulation $\mathbf{x}'$ such that $\mathbf{c}'\mathbf{x}'$ is a minimum is called an optimal circulation. The last component of any optimal circulation is the maximum flow value in G, which is assumed to be finite. So $\mathbf{c}'\mathbf{x}'$ is finite for any circulation $\mathbf{x}'$. Thus the upper-bounded transshipment problem to obtain an optimal circulation in G' is a feasible problem and cannot be unbounded.

Hence there exists a circulation $\mathbf{x}'$ which is a feasible tree solution in G'. Now the last component of $\mathbf{x}'$ is the flow along the arc e from n to i, which is the maximum flow value v, assumed to be finite. The capacity of this arc is infinite. So this arc e can never be saturated. As before, we use the arcs of the tree and the cost vector $\mathbf{c}$ to obtain the vector $\mathbf{y}$. If e is in the tree we have $y_n - 1 = y_1$. If e is not in the tree it has to be a free arc which cannot be profitable because $\mathbf{x}'$ is optimal. In that case $y_n - 1 \geqslant y_1$. Thus in any case $y_n > y_1$, which implies $S = \{i : y_i \leqslant y_1\}$ and $T = \{i : y_i > y_1\}$ are proper subsets of V. Thus (S, T) is a cut in the network.

Let (i, j) be a forward arc in (S, T). Then $y_i + c_{ij} = y_i \leqslant y_1$ and $y_j > y_1$. So $y_i + c_{ij} < y_j$ which implies (i, j) is not in the tree. But this is precisely the condition for a free arc (i, j) to be profitable. And there are no profitable arcs since $\mathbf{x}'$ is optimal. Thus every forward arc in (S, T) is saturated, proving property 1.

Next let (i, j) be a backward arc in (S, T). This time $y_i + c_{ij} > y_j$, which implies (i, j) is not in the tree. But this is precisely the condition for a saturated arc to be profitable. Thus every backward arc is free, proving property 2.

Now $v = \Sigma\Sigma(x_{ij} - x_{ji})$, where the summation is over $i \in S$, $j \in T$. Every forward arc is full, implying $x_{ij} = u_{ij}$ in this summation. Every backward arc is free, implying $x_{ji} = 0$ in this summation. Hence $v = \Sigma\Sigma u_{ij}$, where $i \in S$ and $j \in T$. In other words, $v = C(S, T)$, proving property 3. This completes the proof.

Flow augmenting paths

Earlier in this section we saw that the maximum flow problem can be solved as an upper-bounded transshipment problem. What makes the theory of the MFP interesting is the fact there are some efficient algorithms for solving it, and we will be discussing briefly the development of some of these algorithms in this section and the next.

Let G be an upper-bounded, weakly connected, directed graph with a unique source and a unique sink. Take a simple path between the source and the sink in the underlying (undirected) graph of G. After taking this path reintroduce the direction in each edge of the path so that each edge becomes an arc. The digraph thus obtained from the path is an upper-bounded directed graph P which is called a **simple path** from source to sink in the network. An arc in P directed towards the sink is a forward arc, otherwise it is a backward arc. A path P from source to sink is called a **flow augmenting path** (FAP) with respect to a feasible flow if no forward arc in P is saturated and no backward arc in P is free.

If P is an FAP with respect to a feasible flow $\mathbf{x}$, then each forward arc in P has a **positive excess capacity** which is the difference between the capacity of the arc and the current flow in the arc. The minimum of the excess capacities of all the forward arcs and the current flows of all the backward arcs is a positive number t. Now we increase the flow in each forward arc by t units and decrease the flow in each backward arc by t units, resulting in an updated flow $\mathbf{x}'$ in G. Notice that $\mathbf{x}'$ is a feasible flow in G because (i) vertices not in P are unaffected; (ii) each intermediate vertex in P received t more units and sent out t more units without violating any capacity constraints; (iii) the source sent out t more units without violating any capacity constraints, and (iv) the sink received t more units without violating any constraints.

We also observe that after sending t units like this, P is not an FAP with respect to $\mathbf{x}'$ because at least one forward arc becomes saturated or at least one backward arc becomes free. The conclusion is that if $\mathbf{x}$ is a feasible flow with flow value v and if P is an FAP with respect to $\mathbf{x}$, then we can use P to obtain a feasible flow $\mathbf{x}'$ with flow value $v + t$, where t is positive. In other words, if $\mathbf{x}$ is an optimal flow then there is no FAP with respect to $\mathbf{x}$. The pleasant fact is that the converse also is true and so we have the following result.

Theorem 4.5
A feasible flow is optimal if and only if there is no flow augmenting path with respect to it.

Proof
It is already established that if the current flow value is optimal, then there is no FAP with respect to this flow.

Suppose $\mathbf{x}$ is a flow with a flow value v such that there is no FAP with respect to $\mathbf{x}$. Let $G = (V, E)$, where $V = \{1, 2, \ldots, n\}$. As usual, the source is vertex 1 and the sink is vertex n. Let S be the set of vertices i such that there is an FAP from the source to vertex i. Then $1 \in S$. Let $T = V - S$. Then $n \in T$ because by hypothesis there is no FAP from the source to the sink. Thus we have a cut (S, T) in G.

Suppose (i, j) is an arc in E such that $i \in S$ and $j \in T$. If the flow in this arc is less than its capacity then we will have an FAP from 1 to j, which is a contradiction. So every forward arc in the cut (S, T) is saturated. Likewise, we can prove that every backward arc in (S, T) is free. Hence $v = C(S, T)$. Thus we have produced a cut such that the current flow value equals the cut value of this cut, showing that the current flow is indeed optimal. This completes the proof.

As a result of this theorem we can solve the MFP by examining flow augmenting paths one by one and updating the flow at each stage. If all the components of the capacity vector are integers, then, at each iteration, t is a positive integer and the value v of the flow increases by t. No flow value can exceed the positive integer $C(S, T)$, where $S = \{1\}$ and $T = V - S$. Thus the value v cannot keep increasing by an integer amount more than a finite number of times. So there will be a stage at which no FAP exists, giving an optimal flow after a finite number of iterations. If the components of the capacity vector are rational numbers, then let r be the common denominator of all these components. We then solve the MFP with the capacity vector $r\mathbf{u}$

whose components are all integers. So we assert that if all the capacities are rational numbers then this procedure will definitely solve the MFP. In general, if some capacities are not rational numbers then the procedure need not terminate after a finite number of iterations. See Lawler (1976) and Papadimitriou and Steiglitz (1982) for details.

The labeling algorithm

In order to solve the MFP using flow augmenting paths, we need a systematic procedure for finding a flow augmenting path whenever it exists. We do this by propagating labels from the source until we reach the sink or get stuck. A vertex is a **labeled vertex** if there is an FAP from the source to that vertex. The **label** $L(i)$ of a labeled vertex is a pair $[t, k]$ of two numbers in which t is the extra flow that the vertex i can obtain from the source using the FAP under consideration, and (k, i) or (i, k) is the last arc in this path. For example, if 3 is a labeled vertex with label $[9, 5]$ then there is an FAP from the source to vertex 3 in which the last arc is $(5, 3)$ or $(3, 5)$, and using this path we can send 9 more units of flow from the source to vertex 3. So if the sink can be labeled the existing flow is not optimal. The converse also is true: if the sink cannot be labeled then the current flow is optimal. We state this fact as a theorem.

Theorem 4.6
The current flow value is optimal if and only if the sink has no label.

Proof
If the flow value is optimal, the sink cannot have a label.

On the other hand, suppose the sink n has no label. Let S be the set of labeled vertices and T the set of unlabeled vertices. Then the source 1 is in S and the sink n is in T and (S, T) is a cut. Obviously, every forward arc in the cut is saturated and every backward arc is free, implying v is the cut value of this cut. Thus for the existing flow with flow value v there is a cut (S, T) such that $v = c(S, T)$. This completes the proof.

If we adopt this labeling procedure and choose flow augmenting paths, we will ultimately obtain the optimal value. However, if we do not have a judicious procedure for choosing the flow augmenting paths, it is possible that the number of iterations needed to obtain the optimal flow could be as large as the maximum flow value v in the worst case. This is illustrated by the following example.

Example 4.5
In the network shown below, the capacities of all arcs except $(2, 3)$ are M and the capacity of $(2, 3)$ is 1.

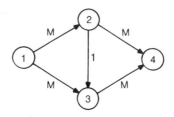

It is easy to see that we can send M units from source 1 to sink 4 along $1 \to 2 \to 4$ and then M units along $1 \to 3 \to 4$. At this stage the flow is optimal, with maximum flow value equal to $2M$, and the number of iterations needed is only 2. But it is possible to solve this problem after performing M iterations as shown below. (And if M is large we are in serious trouble!)

The initial flow is $\mathbf{x} = \mathbf{0}$.

Iteration 1
Consider the path $1 \to 2 \to 3 \to 4$ with flow 0 in each arc. This is an FAP with $t = 1$. Send 1 unit along this path. Then the forward arc $(2, 3)$ becomes saturated. The current flow value is $v = 1$.

Iteration 2
Consider the path $1 \to 3 \leftarrow 2 \to 4$ with flow 0 in the forward arc and flow 1 in the backward arc. This is an FAP with $t = 1$. Send 1 unit along this path. The current flow value is $v = 2$.

Iteration 3
Consider the path $1 \to 2 \to 3 \to 4$ with flow 0 along $(2, 3)$ and flow 1 along the other arcs. This is an FAP for which $t = 1$. Send 1 unit along this path. The current flow value is $v = 3$.

If we continue like this, it will take M iterations before we obtain the maximum flow value.

Observe that in this example the number of arcs in an FAP in the case involving two iterations is less than the number of arcs in an FAP involving $2M$ iterations. On the other hand, if we interchange the capacities of $(1, 3)$ and $(2, 3)$ in the network of Example 4.5, then the optimal flow can be obtained after two iterations whether we take FAPs with two arcs or three arcs. However, it looks as though the efficiency of this procedure depends upon the number of arcs in the FAP we choose. This is indeed the case, as stated in the following path-breaking theorem. For a proof, the reader may refer to Lawler (1976).

Theorem 4.7
If each flow augmentation is made along an augmenting path with a minimum number of arcs, the number of iterations does not exceed $mn/2$, where m is the number of arcs and n is the number of vertices in the network.

We now discuss the shortest augmenting path algorithm based on this theorem in the remainder of this section.

Edmonds–Karp labeling

Let $\mathbf{x}$ be a feasible flow in $G = (V, E)$. Initially $\mathbf{x} = \mathbf{0}$. We construct a directed network $G(\mathbf{x})$ as follows:

1. If (i, j) is a free arc in G, then we have an arc from i to j in $G(\mathbf{x})$ with capacity equal to the capacity of (i, j).
2. If (i, j) is a saturated arc in G, then we have an arc from j to i in $G(\mathbf{x})$ with capacity equal to that of (i, j).
3. If (i, j) is an arc in G which is neither free nor saturated, then we construct two arcs – one from i to j with capacity equal to $u_{ij} - x_{ij}$ and the other from j to i with capacity x_{ij}. Notice that if both (p, q) and (q, p) are edges in G which are neither free nor saturated we will have two arcs from p to q and two arcs from q to p in $G(\mathbf{x})$. We can avoid the possibility of facing multiple arcs of this sort if we make a tacit assumption that if (p, q) is an arc in G then there is no arc from q to p in G. Or if both (p, q) and (q, p) are arcs in G, then we introduce an extra vertex r in one of the two arcs. If r is introduced as an extra vertex in (p, q) between p and q, splitting the arc (p, q) into two arcs (p, r) and (r, q), then we assign the capacity of (p, q) to both (p, r) and (r, q), creating a directed network $G' = (V', E')$ in which both (p, q) and (q, p) cannot be arcs at the same time. When all is said and done it is possible to construct a network $G(x)$ without multiple arcs. It is a simple exercise to verify that a flow augmenting path with respect to the existing flow $\mathbf{x}$ exists in G if and only if there is a directed path in $G(\mathbf{x})$ from the source to the sink.

In particular, a directed path from source to sink in $G(\mathbf{x})$ with k arcs corresponds to an FAP in G with k arcs. Thus the problem of finding an FAP in G with as few arcs as possible is equivalent to the problem of finding a directed path in $G(\mathbf{x})$ from source to sink with as few arcs as possible. A directed path in $G(\mathbf{x})$ with a minimum number of arcs can be obtained by a breadth-first search (BFS) technique. Figure 4.1 shows the flow chart for this labeling algorithm. Each iteration in this labeling procedure has five steps.

Step 1. Use the current flow $\mathbf{x}$ to display G, with each arc showing both the flow and the capacity.

Step 2. Construct the auxiliary network $G(\mathbf{x})$.

Step 3. Construct the BFS tree rooted at the source in $G(\mathbf{x})$ with labeled and unlabeled vertices. Stop if the sink cannot be labeled, in which case $\mathbf{x}$ is optimal. Otherwise go to step 4.

Step 4. Construct an FAP.

Step 5. Update the flow and go to step 1.

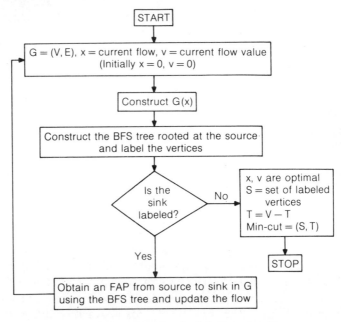

Fig. 4.1 Flow chart for the labeling algorithm.

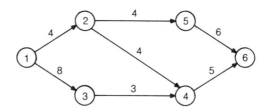

Fig. 4.2 Network discussed in Example 4.6.

The process continues till the sink becomes unlabeled and at this stage a minimum cut (S, T) is obtained, where S is the set of labeled vertices in the last iteration and $T = V - S$.

Example 4.6

Now consider the network shown in Fig. 4.2, where $\mathbf{x}^T = [x_{12} \; x_{13} \; x_{24} \; x_{25} \; x_{34} \; x_{46} \; x_{56}]$ and $\mathbf{u} = [4 \; 8 \; 4 \; 4 \; 3 \; 5 \; 6]$. The source is 1, the sink is 6 and initially $\mathbf{x} = \mathbf{0}$. The procedure requires four iterations, shown in Figs 4.3–4.6.

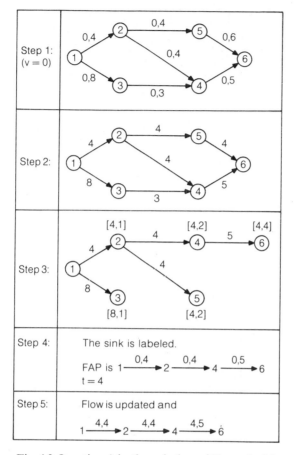

Fig. 4.3 Iteration 1 in the solution of Example 4.6.

4.4 THE MPM ALGORITHM

In this section we discuss a procedure for solving the maximum flow problem in which the flow is augmented not along a single path but along several paths simultaneously. This algorithm, noted for its sophisticated simplicity, is due to V. M. Malhotra, M. Pramodhkumar and S. N. Maheswari, and so may be called the **MPM algorithm**. For a proof of the algorithm, the reader is referred to Papadimitriou and Steiglitz (1982).

Layered networks and cores

Let $G = (V, E)$ be a network as usual, with $V = \{1, 2, \ldots, n\}$, in which the source is at 1 and the sink is at n. Let $\mathbf{x}$ be a feasible flow in G and let $G' = G(\mathbf{x})$ be the network constructed from G using the flow $\mathbf{x}$ as described in section 4.3. The

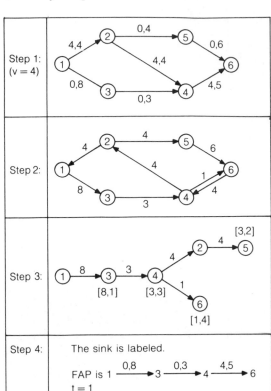

Fig. 4.4 Iteration 2 in the solution of Example 4.6.

weight of each arc in $G(\mathbf{x})$ is its **augmenting capacity**, which may vary as $\mathbf{x}$ varies. Let V_k be the set of vertices in $G(\mathbf{x})$ which can be reached from the source using k arcs in a breadth-first search originating from the source. The set V_k is the **kth layer** corresponding to the given flow $\mathbf{x}$. The subgraph $G'' = (V', E')$ obtained from G', where V' is the union of all its layers and E' has as many arcs as possible with the restriction that every arc in E' is from a lower layer to a higher layer, is called a **layered network** of G with respect to the flow $\mathbf{x}$. Observe that we need a feasible flow to layer a network.

If the feasible flow is not optimal, there is a layer V_r such that the sink is in that layer. Since the aim is to increase the flow to the source using a flow augmenting path with as few arcs as possible in a nonoptimal situation, all layers higher than that of the sink can be ignored. We can also delete any vertex (other than n) in the layered network whose outdegree is 0. The subgraph obtained after these deletions is the **core** of G'' and is denoted $\bar{G}$. So

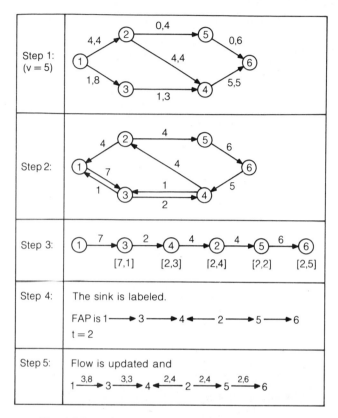

Step 1: (v = 5)	
Step 2:	
Step 3:	
Step 4:	The sink is labeled. FAP is 1 ⟶ 3 ⟶ 4 ⟵ 2 ⟶ 5 ⟶ 6 t = 2
Step 5:	Flow is updated and 1 —3,8→ 3 —3,3→ 4 —2,4← 2 —2,4→ 5 —2,6→ 6

Fig. 4.5 Iteration 3 in the solution of Example 4.6.

if the flow is not optimal, we can construct a core in which there will be at least one directed path from source to sink.

The following diagram shows a network $G(\mathbf{x})$ for a certain feasible flow $\mathbf{x}$ in the network.

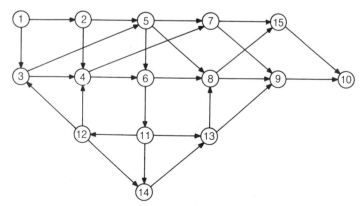

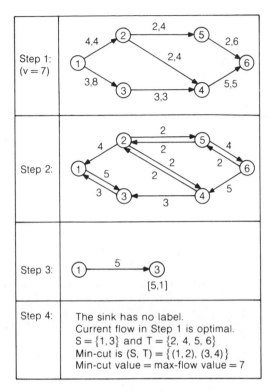

Fig. 4.6 Iteration 4 in the solution of Example 4.6.

The temporary capacities of the arcs in $G(\mathbf{x})$ are not explicitly specified. The BFS technique, starting from vertex 1, gives the layered network as shown below.

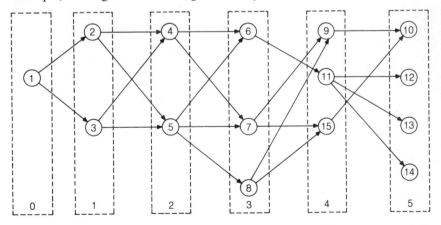

There are six layers and the sink (vertex 15) is in the fourth layer, since the layer number of the source is taken as 0. The fifth layer can be deleted. It is easy to see that arcs (6, 11), (4, 6), (5, 6), (9,10), (7, 9) and (8, 9) can also be deleted from the layered network. The following diagram shows the core, in which there is a directed path from the source to the sink.

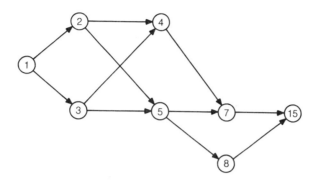

The main part of the algorithm is a clever way of finding an optimal flow in the core. Instead of augmenting the flow along flow augmenting paths one at a time as in section 4.3, we try to send as much flow as possible from source to sink in the core along several paths till the core is no longer useful. Then we construct another core if it exists and repeat the process till the sink becomes unreachable in the BFS search.

Let i be a vertex (which is neither the source nor the sink) in the core corresponding to the current flow $\mathbf{x}$ in the network. If $O(i)$ is the sum of the augmenting capacities of all arcs that start from i and $I(i)$ is the sum of the augmenting capacities of all arcs that end in i, the **potential** $p(i)$ at i is the minimum of $O(i)$ and $I(i)$. The potential of the source is the sum of the auxiliary capacities of all the arcs in the core directed from the source. The potential of the sink is the sum of the augmenting capacities of all the arcs in the core directed to the sink.

An arc (i, j) in the core is **unsaturated** with respect to the flow if the component of the flow along the arc is positive but less than its augmenting capacity. A vertex (other than the sink) is said to be **processed** if the amount of flow that goes through i (for the particular feasible flow $\mathbf{x}$) is distributed along the arcs that start from i in such way that among these arcs there is at most one unsaturated arc. This processing concept is at the heart of the MPM procedure.

A flow diagram for the MPM algorithm appears in Fig. 4.7. Iteration k of the algorithm proceeds as follows:

Step 1. The current feasible flow vector is $\mathbf{x}^k$ in the network. (Initially $\mathbf{x}^k = \mathbf{0}$.) Use $\mathbf{x}^k$ to define the digraph $G' = G(\mathbf{x}^k)$.
Step 2. Use a breadth-first search technique starting from the sink in $G(\mathbf{x}^k)$ to construct a layered network G''. If there is no directed path in G'' from source to sink, go to step 7.

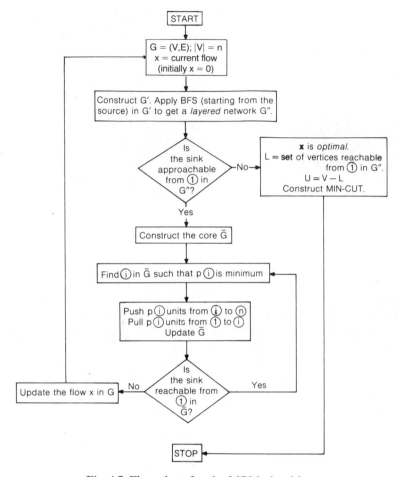

Fig. 4.7 Flow chart for the MPM algorithm.

Step 3. Construct the core from G''.

Step 4. Choose a vertex i of smallest potential in the core.

(a) Push $p(i)$ units of flow from i to the sink in the core, processing vertices en route so that no vertex has to be processed more than once.

(b) Pull $p(i)$ units of flow from the source to i by processing vertices en route as in (a) above.

Step 5. If the sink is reachable from the source in the core, go to step 4. Otherwise go to step 6.

Step 6. Update the flow in G and go to step 1.

Step 7. The algorithm terminates. The current flow is optimal. Let L be the set of vertices reachable from the source and U be its complement. The minimum cut is (L, U).

Notice that (i) in iteration k all flow augmenting paths have the same number of arcs and (ii) the number of arcs in a flow augmenting path in iteration k is less than the number of paths in any flow augmenting path in a subsequent iteration.

A typical maximum flow problem can thus be replaced by several (at most n, where n is the number of vertices) problems, each much easier than the original one, by this layering technique.

Example 4.7
Obtain the maximum flow and a minimum cut in the network shown below.

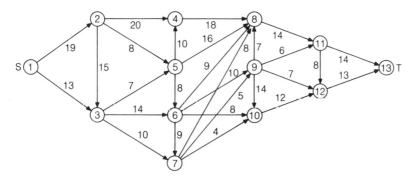

Observe that the total capacity of the outward arcs from the source (vertex 1) is 32 and the total capacity of the inward arcs to the sink (vertex 13) is 27. So the maximum flow value cannot exceed 27.

Iteration 1
The current flow is $\mathbf{x} = \mathbf{0}$ and the current flow value is $v = 0$. The layered graph and the core are as shown in the following diagram:

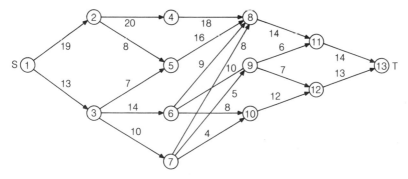

In the core each FAP will have five arcs since the sink is in the fifth layer. Vertex 7 has the smallest potential, with $p(7) = 10$. So we send 10 units from the source to the sink using vertex 7 as shown below:

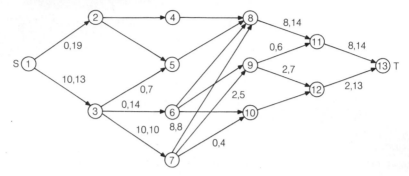

Notice that at every vertex en route, among the forward arcs there should be at most one unsaturated arc, as dictated by the MPM process. Thus we send eight units along $1 \to 3 \to 7 \to 8 \to 11 \to 13$ and two units along $1 \to 3 \to 7 \to 9 \to 12 \to 13$.

At this stage we update the core by deleting all saturated arcs, deleting all vertices (other than the source) with no incoming arcs, and replacing the capacity of each arc by the difference of the current capacity and the current flow. The updated core is as shown below.

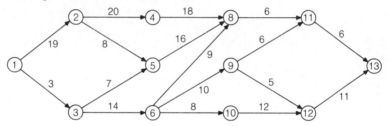

Vertex 3 has the smallest potential with $p(3) = 3$. So we send three units along $1 \to 3 \to 6 \to 9 \to 12 \to 13$. The flow in the updated core is shown in the following diagram, in which each vertex en route is processed.

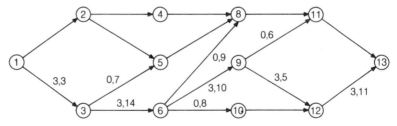

At this stage, the core is updated again as shown below.

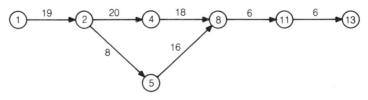

The smallest potential is at both vertex 11 and vertex 13 with $p(11) = p(13) = 6$. So we send six units along $1 \to 2 \to 4 \to 8 \to 11 \to 13$. The flow in the core is as follows.

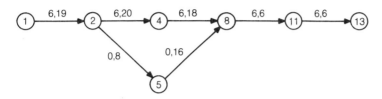

At this stage, if the core is updated, the sink is not reachable from the vertex using the core. The current core is no longer useful. The flow in the given network is called the **blocking flow** when the core ceases to be useful. The blocking flow value is $8 + 2 + 3 + 6 = 19$, and this flow was sent from source to sink along four flow augmenting paths each of length 5. At this stage iteration 1 ends.

Iteration 2
The current flow is as displayed below, with flow value $v = 19$.

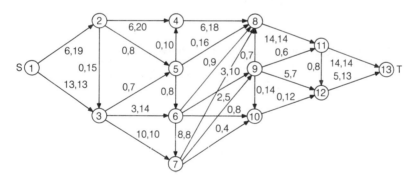

Here is the layered network:

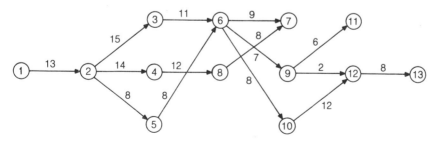

The core is as follows. Notice that the length of any path from source to sink in the core in this iteration is greater than that of the corresponding path in the previous iteration, as it should be.

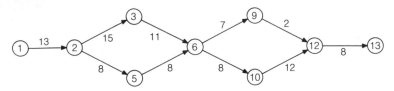

The vertex of minimum potential is at 9 with $p(9) = 2$. So we send two units along $1 \to 2 \to 3 \to 6 \to 9 \to 12 \to 13$. The core is updated as shown below.

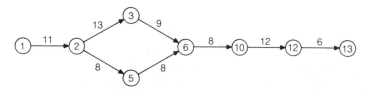

The smallest potential is at vertex 13 with $p(13) = 6$. We send six units along $1 \to 2 \to 3 \to 6 \to 10 \to 12 \to 13$. At this stage the core is no longer useful. So at the end of this iteration, the flow value is $19 + 2 + 6 = 27$.

Iteration 3
The current flow is as shown below, with a flow value of 27:

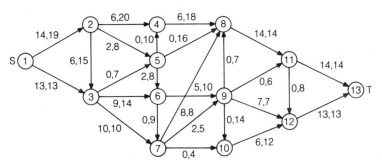

Here is the layered network:

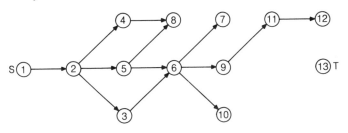

There is no layer to which the sink belongs. So the current flow (as shown at the top of page 150) is an optimal solution for the maximum flow problem in our network. Notice that every vertex (other than the sink) belongs to some layer in the optimal layered network. Thus $L = V - \{13\}$ and $U = \{13\}$. The cut $(L, U) = \{\{11, 13\}, \{12, 13\}\}$, with cut value $14 + 13 = 27$, is a minimum cut. Observe also that in this minimum cut, every forward arc is saturated. There are no backward arcs in this cut. The flow along any backward arc (if it exists) is necessarily zero.

4.5 THE MFP IN PLANAR UNDIRECTED NETWORKS

An undirected graph is said to be **planar** if it is possible to draw it in a plane such that no two edges of the graph intersect except possibly at a vertex. In this section we consider planar undirected networks. Each edge has a capacity which is an upper bound for the permissible flow along that edge in either direction.

A planar undirected network has some special properties enabling us to obtain a minimum cut and a maximum flow vector by solving the (relatively easier) problem of finding a shortest path between two specified vertices in an associated planar network. So we first investigate some properties of such networks.

A **face** of a planar network is a region of the plane bounded by one or more edges such that any two points in the region can be joined by a curve that does not meet any edge or a vertex. A face that is not finite is called an **unbounded face**. The set of edges that enclose a face is the **boundary** of the face. Each edge obviously belongs to the boundary of at most two faces.

Suppose $G = (V, E)$ is a planar simple connected undirected graph where $V = \{1, 2, \ldots, n\}$. The **dual graph** $G^* = (V^*, E^*)$ is constructed as follows. Each face f of G corresponds to a vertex f^* in G^*. If $\{i, j\}$ is an edge of G which is common to the boundaries of two faces f and g, then define an edge $\{f^*, g^*\}$ in E^*. If $\{i, j\}$ is an edge in the boundary of only one face f, we have a loop $\{f^*, f^*\}$ in E^*. Figure 4.8(i) shows the graph G with face a bounded by four edges, face b bounded by three edges and the unbounded face c. Figure 4.8(ii) shows G^*, the dual of G, with vertices a, b and c. The edge $\{2, 6\}$ corresponds to the loop $\{a, a\}$ in G^*. As there are three edges in common to the boundaries of the faces a and c in G, there are three edges between a and c in G^*. The network G^* thus constructed from G is also planar and its dual is G. Thus each is the dual of the other. For convenience, we may designate G as the **primal network** and G^* as the **dual network**. The number of faces in one is equal to the number of vertices in the other. Both G and G^* have the same number of edges. It is also easy to verify that the set of edges in a cycle in a planar network corresponds to the set of edges in a cut in its dual and vice versa. For example, the edges in the cycle 2—5—3—2 in the primal network G of Fig. 4.8 correspond to the edges in the cut $\{\{b, c\}, \{b, c\}, \{a, b\}\}$ in the dual network G^*.

An undirected planar graph with a specified vertex s as the source and a specified vertex t as the sink is called an **s–t planar network** if both s and t are

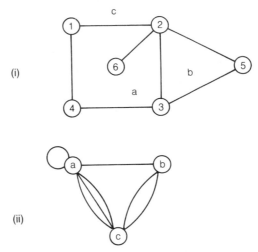

(i)

(ii)

Fig. 4.8 (i) A planar simple connected undirected graph G; and (ii) its dual G'.

vertices in the boundary of the unbounded face. The graph in Fig. 4.8(i) is $s{-}t$ planar if $s=1$ and $t=5$. But it is not an $s{-}t$ planar network if $s=1$ and $t=6$.

Let G be an $s{-}t$ planar network. Draw a new edge joining s and t such that this edge is completely in the unbounded face of G, thus creating an additional face without violating the planarity of the graph. In Fig. 4.9(i) we have an $s{-}t$ network G, in which $s=1$ and $t=6$. The extra edge is the dashed line joining vertex 1 and vertex 6.

We now construct the dual of this enlarged $s{-}t$ graph in which the additional face corresponds to a specified vertex s^* and the unbounded face corresponds to a specified vertex t^*. In this dual graph, the edge joining s^* and t^* is deleted. The resulting planar graph G^* is the $s{-}t$ **dual** of G (Fig. 4.9(ii)).

It is obvious that any path between s^* and t^* in G^* is an $s{-}t$ cut in G and vice versa. By definition, the capacity of an edge between i^* and j^* in G^* is the capacity of an edge in G which is common to the sets of boundaries of the faces which correspond to i^* and j^*. Consequently, the capacity of an $s{-}t$ cut in G is equal to the length of the corresponding path between s^* and t^*. Thus a shortest path between s^* and t^* in the dual network will define a minimum $s{-}t$ cut in the given network. Consequently, the cut value of this minimum cut will give the flow value of a maximum flow in the given network. What is not immediately obvious is the fact that any shortest path between s^* and t^* in the dual network will also define an optimal flow in the primal network. This is the subject of our next theorem.

Theorem 4.8
Let G be an $s{-}t$ network and G^* its dual. Let $d(k^*)$ be the shortest distance between s^* and k^* in G^*. Define $x_{ij} = d(j^*) - d(i^*)$, where $\{i^*, j^*\}$ is the edge which corresponds to $\{i, j\}$ in G. Then $\mathbf{x} = [x_{ij}]$ is an optimal flow in G.

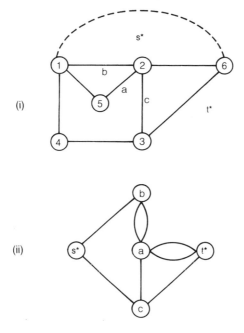

Fig. 4.9 (i) An s–t network; and (ii) its dual.

Proof

According to the definition, if $x_{ij} = d(j^*) - d(i^*)$ is positive, then x_{ji} is negative. Since the network is undirected we can regard the negative flow from j to i as positive flow from i to j. Hence without loss of generality the flow vector thus defined can be considered as a nonnegative vector. Now

$$d(j^*) \leqslant d(i^*) + [\text{capacity of } \{i^*, j^*\}]$$

So

$$x_{ij} = d(j^*) - d(i^*) \leqslant [\text{capacity of } \{i^*, j^*\}]$$

Thus $\mathbf{x}$ is a flow in a capacitated network G.

Let k be any vertex in G other than the source or the sink. Consider the cut $(\{k\}, V - \{k\})$ in G. The edges in this cut correspond to a cycle C^* in the dual network. Thus the netflow at the vertex k is the sum $\Sigma(d(j^*) - d(i^*))$ over the edges in the cycle. The terms in this sum cancel each other out. Thus the netflow at each vertex other than the source and the sink is 0.

Suppose P^* is a shortest path between s^* and t^* in the dual network. Now

$$d(j^*) = d(i^*) + [\text{capacity of } \{i^*, j^*\}]$$

for any edge $(i^*, j^*\}$ in P^*, which implies the edge $\{i, j\}$ is saturated. In other words, the flow $\mathbf{x}$ thus defined saturates all the edges in the shortest path P^*. So it is an optimal flow.

Example 4.8

Find a maximum flow vector and a minimum cut in the s–t planar undirected network $G = (V, E)$ shown below. The source is at vertex 1 and the sink is at vertex 4. The capacity of each edge is given on the edge. The finite faces are a and b.

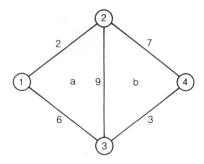

Draw the extra edge joining the source and the sink which lies completely in the unbounded face of G, giving the additional face s^* which corresponds to the source vertex in the dual network, as shown in the diagram below. The unbounded face in the enlarged network is t^*, which is the sink vertex in G^*. Both the dual and the primal are drawn in the same diagram and the edges in the dual network are dashed lines. This enables us to identify by inspection the capacities of corresponding edges. For example, the capacity of the edge between s^* and a is the capacity of the edge between 1 and 3 in the primal network because the edge $\{1, 3\}$ and the edge $\{s^*, a\}$ cross each other exactly once, identifying their correspondence.

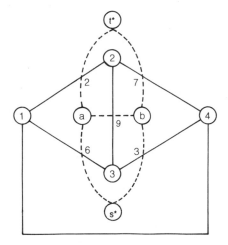

The shortest distance between s^* and t^* is 8, which is the minimum cut value in G. The edge $\{1, 2\}$ corresponds to $\{a, t^*\}$. So

$$x_{12} = d(t^*) - d(a) = 8 - 6 = 2$$
$$x_{13} = d(a) - d(s^*) = 6 - 0 = 6$$
$$x_{24} = d(t^*) - d(b) = 8 - 3 = 5$$
$$x_{23} = d(b) - d(a) = 3 - 6 = -3$$
$$x_{34} = d(b) - d(s^*) = 3$$

(note that we would have $x_{23} = x_{32} = 3$ in an undirected graph). The optimal flow vector is as displayed below.

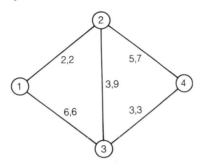

4.6 MULTITERMINAL MAXIMUM FLOWS

Unlike the s–t maximum flow problem of sections 4.3–4.5 in the context of moving as much as possible of a single commodity from a unique source s to a unique sink t in a capacitated network $G = (V, E)$, in this section every vertex is considered as both a source and a sink. Thus we are interested in obtaining the maximum flow value v_{ij} from vertex i to vertex j for every i and j in $V = \{1, 2, \ldots, n\}$. For a fixed pair of vertices, when we compute the maximum flow value from one to the other, the other $n - 2$ vertices are considered as intermediate vertices. Thus this all-pair maximum flow problem can be handled by solving $n(n - 1)$ distinct maximum flow problems in the case of a directed network and $n(n - 1)/2$ maximum flow problems in the undirected case. We now discuss a method of finding the maximum flow between every pair of vertices in an undirected graph in which an optimal solution is obtained after solving $(n - 1)$ problems instead of $n(n - 1)/2$ problems.

Let $f_{ij} = f_{ji}$ be the maximum flow value between i and j in a capacitated network, where i and j are two distinct vertices. We assume that, for each i, f_{ii} is $+\infty$, usually denoted by a large positive number M. The $n \times n$ symmetric matrix $\mathbf{F} = [f_{ij}]$ is the **flow matrix** of the network. If the graph is a tree, then f_{ij} is the weight of the edge of minimum weight in the unique path joining i and j.

In this section we study two problems. One is network realizability, where the object is to obtain a necessary and sufficient condition to be satisfied by an arbitrary symmetric matrix $\mathbf{F}$ so that each nondiagonal entry f_{ij} in the matrix is the maximum flow value between the vertices i and j of an undirected network. The other is network analysis, where obtaining the maximum flow value and the minimum cut between every pair of vertices in an undirected network is of interest.

Theorem 4.9 (Gomory–Hu realizability theorem)

A necessary and sufficient condition for the symmetric matrix $[f_{ij}]$ to be the flow matrix of an undirected graph is $f_{ik} \geqslant \min\{f_{ij}, f_{jk}\}$ for all i, j and k.

Proof

Suppose f_{ik} is the maximum flow value between i and k in an undirected graph. So there exists a cut (S, T) such that f_{ik} is the cut value $C(S, T)$ of this cut.

Assume $i \in S$. Then $k \in T$. Let j be another vertex. If $j \in S$, then (S, T) is a cut separating j and k which implies $f_{jk} \leqslant C(S, T)$. On the other hand, if $j \in T$, the same (S, T) is a cut separating i and j, which implies $f_{ij} \leqslant C(S, T)$. Thus

$$\min\{f_{ij}, f_{jk}\} \leqslant C(S, T) = f_{ik}$$

So the condition is necessary.

Suppose we are given a matrix $\mathbf{F} = [f_{ij}]$ which satisfies the 'triangle inequality' $f_{ik} \geqslant \min\{f_{ik}, f_{kj}\}$ for all i, j and k. Then by a simple inductive argument it can be shown that

$$f_{in} \geqslant \min\{f_{12}, f_{23}, f_{34}, \ldots, f_{n-1,n}\}$$

Let T be a maximum weight spanning tree in the complete network K_n in which the weight of the edge $\{i, j\}$ is f_{ij}. Suppose the unique path between i and j in the tree is

$$i-1-2-3-\cdots-p-j$$

Let h_{ij} be the maximum flow value between i and j in T. Then

$$h_{ij} = \min\{f_{i1}, f_{12}, f_{23}, \ldots, f_{pj}\} = f_{ij}$$

since the edge $\{i, j\}$ is not an edge in the maximum weight spanning tree. Thus corresponding to the given matrix $[f_{ij}]$, there exists a graph T in which the maximum flow value between i and j is f_{ij}.

Example 4.9

Construct a tree $T = (V, E)$, where $V = \{1, 2, 3, 4, 5, 6\}$, in which the maximum flow between i and j is the (i, j)th element in the following 6×6 symmetric matrix:

$$\mathbf{F} = \begin{bmatrix} M & 25 & 25 & 25 & 25 & 20 \\ 25 & M & 32 & 32 & 29 & 20 \\ 25 & 32 & M & 37 & 29 & 20 \\ 25 & 32 & 37 & M & 29 & 20 \\ 25 & 29 & 29 & 29 & M & 20 \\ 20 & 20 & 20 & 20 & 20 & M \end{bmatrix}$$

The five edges of the maximum weight spanning tree T (obtained by the greedy algorithm) in the complete graph K_6 in which the weight of the edge $\{i, j\}$ is the (i, j)th element in the matrix are $\{1, 2\}, \{1, 6\}, \{2, 3\}, \{3, 4\}$ and $\{2, 5\}$.

Notice that the unique path in this tree between 4 and 5 is 4—3—2—5 in which the edge of minimum weight is $\{2, 5\}$, with weight 29. The $(4, 5)$th entry in $\mathbf{F}$ is also 29.

Observe that in the 6×6 flow matrix of Example 4.9 there are five distinct numbers: 20, 25, 29, 32 and 37. This leads us on to the following theorem.

Theorem 4.10

The number of distinct nondiagonal numbers in an $n \times n$ flow matrix cannot exceed $n-1$.

Proof

Consider a maximum weight spanning tree in the complete graph K_n defined by the flow matrix $\mathbf{F}$ as before. This tree has $n-1$ edges. The claim is that every flow value in $\mathbf{F}$ is equal to the weight of one of these $n-1$ arcs. The proof is by contradiction.

Suppose the flow value f_{ij} is not equal to the weight of any of the edges in T for a fixed i and j. Let the path in T between i and j be i—1—2—p—j. Suppose $\{u, v\}$ is an edge of minimum weight in this path. Then $f_{ij} \geqslant f_{uv}$. If $f_{ij} > f_{uv}$, we can replace the edge $\{u, v\}$ by $\{i, j\}$ and obtain a spanning tree whose weight exceeds the weight of T, which is a contradiction.

The Gomory–Hu tree

Suppose $\mathbf{F}$ is the flow matrix of an undirected graph. As we saw earlier, it is possible to construct a tree T (not necessarily unique) which also has the same flow matrix. The weight of the edge $\{i, j\}$ of minimum weight in the unique path (in the tree) joining two vertices u and v is, of course, the minimum cut value corresponding to the maximum flow value between u and v. But if we partition the set of vertices of the tree into two sets S and T by deleting the edge $\{i, j\}$, then the cut (S, T) need not be a minimum cut associated with the maximum flow value between the vertices u and v. Here is an example to show that this possibility cannot be ruled out.

Example 4.10

Consider the undirected graph G as shown below. The capacity of each edge is given on the edge.

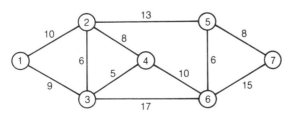

The flow matrix is found to be:

$$F = \begin{bmatrix} M & 19 & 19 & 19 & 19 & 19 & 19 \\ 19 & M & 37 & 23 & 27 & 37 & 23 \\ 19 & 37 & M & 23 & 27 & 37 & 23 \\ 19 & 23 & 23 & M & 23 & 23 & 23 \\ 19 & 27 & 27 & 23 & M & 27 & 23 \\ 19 & 37 & 37 & 23 & 27 & M & 23 \\ 19 & 23 & 23 & 23 & 23 & 23 & M \end{bmatrix}$$

A tree with the same flow matrix is the following:

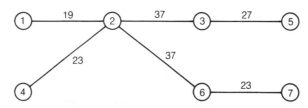

The maximum flow value between 2 and 6 in the graph is 37, which is the weight of edge $\{2, 6\}$. This edge partitions the set of vertices into two sets $S = \{1, 2, 3, 4, 5\}$ and $T = \{6, 7\}$, giving a cut $(S, T) = \{\{3, 6\}, \{4, 6\}, \{5, 6\}, \{5, 7\}\}$ with cut value $17 + 10 + 6 + 8 = 41$, which exceeds the minimum cut value.

In the network analysis problem, we are interested in finding the flow matrix of a given undirected graph by constructing a tree (with the same flow matrix) from which we can obtain the minimum cuts in addition to the maximum flow values. A tree whose flow matrix is the same as that of the network is called a **cut tree** or **Gomory–Hu tree** of the network if each edge of the tree represents a minimum cut of the network.

We now describe briefly a procedure for constructing such a tree (for details see Hu, 1982). The Gomory–Hu theory is based on the following three facts:

1. Suppose S is set of vertices in the network $G = (V, E)$. Then the set S can be condensed into a single vertex to construct a condensed graph. If v is a vertex in $V - S$ with edges connecting v and vertices in S, then these edges are replaced by a single edge joining v and S, with weight equal to the sum of the weights of these edges.
2. If (S, T) is any minimum cut in the network, we can condense the set S into a single vertex if we are interested in finding the maximum flow between two vertices in T. Similarly, we can condense T into a single vertex if we are interested in computing the maximum flow between two vertices in S.
3. If (S, T) is a minimum cut obtained in finding the maximum flow between a vertex v and a vertex w, where $v \in S$ and $w \in T$, then we can condense S into a single vertex if we are interested in finding the maximum flow between

v and any vertex in T. Similarly, we can condense T into a single vertex if we are interested in finding the maximum flow between w and any vertex in S.

The Gomory–Hu procedure can be summarized as follows. Initially we choose two vertices i and j at random, and solve the maximum flow problem for these two vertices. Suppose the maximum flow value is f_{ij} and the minimum cut is (S, T). We construct a tree with two vertices S and T and join them with an edge of weight f_{ij}. To obtain a Gomory–Hu tree we have to solve $n-2$ more problems. Each vertex of the tree is a subset of the set of vertices of the network. At a subsequent stage, select any vertex X (from the tree obtained in the previous problem) which as a set has more than one vertex of the network. Suppose u and v are any two vertices in X. Do a maximum flow computation in the network between these two vertices. Suppose f_{uv} is the maximum flow value and (S', T') is the minimum cut in the network, where $u \in S'$ and $v \in T'$. Now X can be partitioned into two subsets Y and Z such that $Y \subset S'$ and $Z \subset T'$. The vertex X in the tree is split into two vertices Y and Z, and these are joined by an edge with weight f_{uv}. If there is an edge in the tree joining U and X then U is either a subset of S' or a subset of T'. In the former case, U and u are on the same side of the minimum cut separating u and v. We join U and Y by an edge with weight equal to the weight of the edge joining U and X. Otherwise we join U and Z (see Example 4.11 below). After $n-2$ more maximum flow computations, we are left with a tree with n vertices which is the desired tree.

Example 4.11
Find a Gomory–Hu tree for the undirected network in Example 4.10. Since the graph has seven vertices, according to the Gomory–Hu theory, we should be able to construct the desired tree by solving six maximum flow problems.

Maximum flow problem 1
The maximum flow value between two randomly chosen vertices 1 and 7 of the network is 19. The minimum cut is (S, T) where $S = \{1\}$ and $T = \{2, 3, 4, 5, 6, 7\}$.

The tree at this stage, as shown below, has two vertices S and T, and the weight of the edge is 19.

Maximum flow problem 2
Take any vertex X from the tree in the previous problem which has at least two vertices of the network and find a maximum flow between them in the network. We take $X = \{2, 3, 4, 5, 6, 7\}$ and vertices 6 and 7. The maximum flow

value between 6 and 7 is 23. The minimum cut is $S' = \{7\}$ and $T' = \{1, 2, 3, 4, 5, 6\}$. So the vertex X is partitioned into two sets $Y = \{7\}$ and $Z = \{2, 3, 4, 5, 6\}$ such that $Y \subset S'$ and $Z \subset T'$.

At this stage the tree has three vertices: Y, Z and $\{1\}$ (see the diagram below). The weight of the edge joining Y and Z is 23. Both $\{1\}$ and Z are subsets of T'. In other words, both $\{1\}$ and Z are 'on the same side' of the current minimum cut. So we join $\{1\}$ and Z by an edge of weight 19.

Maximum flow problem 3

The tree in the previous problem has a vertex $X = \{2, 3, 4, 5, 6\}$ which has more than one vertex from the network. Take two vertices at random from this set and solve the maximum flow problem in the network. We choose 2 and 6.

The maximum flow value is 37. The minimum cut is (S', T'), where $S' = \{1, 2, 5\}$ and $T' = \{3, 4, 6, 7\}$. The set $X = \{2, 3, 4, 5, 6\}$ is partitioned into two sets $Y = \{2, 5\}$ and $Z = \{3, 4, 6\}$ such that $Y \subset S'$ and $Z \subset T'$. The vertices of the tree are Y, Z, $\{1\}$ and $\{7\}$. We join Y and Z by an edge with weight equal to the current maximum flow value 37. The vertex $\{1\}$ and the vertex Y are on the same side of the minimum cut since they are both subsets of S'. So we join $\{1\}$ and Y by an edge of weight 19. The vertices $\{7\}$ and Z are on the same side of the minimum cut. So we join Z and $\{7\}$ by an edge of weight 23.

The current tree is as shown below.

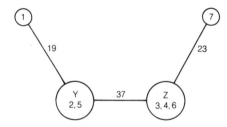

Maximum flow problem 4

$X = \{3, 4, 6\}$. The maximum flow value between 3 and 6 is 37. The minimum cut is (S', T'), where $S' = \{3\}$ and $T' = \{1, 2, 3, 4, 5, 6, 7\}$. So $Y = \{3\}$ and $Z = \{4, 6\}$. The weight of the edge $\{Y, Z\}$ is 37. The vertex $\{2, 5\}$ and the vertex Z are both subsets of T'. So $\{2, 5\}$ and Z are joined by an edge with weight 37. The vertices $\{7\}$ and Z are both subsets of T'. So $\{7\}$ and Z are joined by an edge of weight 23.

The current tree, with vertices Y, Z, $\{2, 5\}$, $\{1\}$ and $\{7\}$, is shown below.

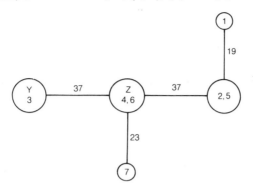

Maximum flow problem 5

$X = \{4, 6\}$. The maximum flow value between 4 and 6 is 23. The minimum cut is $S' = \{4\}$ and $T' = \{1, 2, 3, 5, 6, 7\}$. $Y = \{4\}$ and $Z = \{6\}$. Join Y and Z by an edge of weight 23. The vertices $\{2, 5\}$, $\{3\}$, $\{7\}$ and Z are all subsets of T'. So join $\{2, 5\}$, $\{3\}$, $\{7\}$ and Z with edges of weights 37, 37 and 23 as shown below.

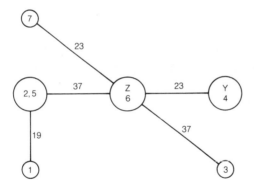

Maximum flow problem 6

$X = \{2, 5\}$. The maximum flow value between 2 and 5 is 27. $S' = \{5\}$ and $T' = \{1, 2, 3, 4, 6, 7\}$. $Y = \{2\}$ and $Z = \{5\}$. The vertices $\{6\}$, $\{1\}$ and Y are subsets of T'. So join $\{6\}$ and Y by an edge of weight 37, and $\{1\}$ and Y by an edge of weight 19.

The current tree, shown below, is a Gomory–Hu tree for the given network since the cardinality of each vertex is 1. This tree is indeed a minimum cut tree. The weight of the edge $\{2, 6\}$ in this tree is 37, which is the weight of an edge of minimum weight in the unique path joining these two vertices in the tree. It is also the maximum flow value between 2 and 6. This edge of minimum weight defines the cut (S, T), where $S = \{1, 2, 5\}$ and $T = \{3, 4, 6, 7\}$, with cut value $6+8+9+6+8 = 37$. So (S, T) is a minimum cut.

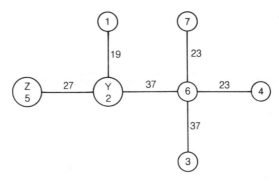

4.7 THE MINIMUM COST FLOW PROBLEM

Any capacitated transshipment problem is considered as a minimum cost flow problem (MCFP) in general. But there is one special type of MCFP which can be designated as 'the' MCFP. Given a capacitated network with a specified vertex as the source, a specified vertex as the sink, a nonnegative cost vector defined on the arcs of the network and a positive integer r, is it possible to ship r units of a commodity from the source to the sink using the arcs of the network? If the answer is yes, what is the minimum cost of doing this? Obviously we can ship r units from the source to the sink if and only if r cannot exceed the maximum flow value. Once this is settled, the problem is like any other capacitated transshipment problem. Thus the MCFP is a combination of the MFP and the capacitated transshipment problem and as such we need not break any new ground to handle it. However, in this section, we examine the problem from a different perspective.

The formal statement of the MCFP is as follows. Let A be the vertex–arc incidence matrix of a network $G = (V, E)$, where $V = \{1, 2, \ldots, n\}$. Vertex 1 is the source and n is the sink. The capacity vector is u, and c is its cost vector. Let $b^T = [-r \; 0 \; 0 \cdots r]$ be the vector with n components where r is a given positive integer which does not exceed the maximum flow value. The MCFP seeks a nonnegative vector x such that (i) $x \leqslant u$; (ii) cx is as small as possible; and (iii) $Ax = b$.

Algorithm cycle

Suppose x is a feasible flow vector for the MCFP with flow value r. This means $0 \leqslant x \leqslant b$ and $Ax = b$. The current cost is $z = cx$. We would like to test whether x is optimal. If it is not, how can we obtain a better solution using the current feasible solution which will eventually lead us to an optimal solution? We now discuss an iterative procedure for accomplishing this.

We assume as before that if (i, j) is an arc in the network G, then (j, i) is not an arc in it. Let $N(\mathbf{x})$ be the network (with the same vertices and arcs as in G) in which each arc (i, j) is associated with the pair (x_{ij}, u_{ij}) where $0 \leqslant x_{ij} \leqslant u_{ij}$. The network $N(\mathbf{x})$ describes the current flow pattern in the network. An arc (i, j) in $N(x)$ is a **free** arc if $x_{ij} = 0$ and it is a **saturated** arc if $x_{ij} = u_{ij}$.

Now we define a weighted network $N'(\mathbf{x})$ as follows. If (i, j) is a free arc in $N(\mathbf{x})$, then (i, j) is an arc in $N'(\mathbf{x})$ with weight c_{ij}. If (i, j) is a saturated arc in $N(\mathbf{x})$, then (j, i) is an arc in $N'(\mathbf{x})$ with weight $-c_{ij}$. If (i, j) is neither free nor saturated, then both (i, j) and (j, i) are arcs in $N'(\mathbf{x})$ with weights c_{ij} and $-c_{ij}$, respectively.

Consider any directed cycle C' in $N'(\mathbf{x})$. The weight w of C' is the sum of the weights of the arcs in C'. If $w < 0$, then C' is called a **negative cycle**. Let C be the corresponding cycle in $N(\mathbf{x})$, which need not be a directed cycle. An arc (p, q) in C is a **positive arc** if (p, q) is an arc in C'. An arc (p, q) is a **negative arc** in C if (q, p) is an arc in C'.

Now let

$$\delta = \min \left\{ \begin{array}{cc} (u_{ij} - x_{ij}), & x_{ij} \\ (i, j) \text{ positive} & (i, j) \text{ negative} \\ \text{in } C & \text{in } C \end{array} \right\}$$

Define a new flow vector $\mathbf{x}'$ in G such that (a) $x'_{ij} = x_{ij} + \delta$ if (i, j) is a positive arc in C, (b) $x'_{ij} = x_{ij} - \delta$ if (i, j) is a negative arc in C and (c) $x'_{ij} = x_{ij}$ if (i, j) is not in C. It is easily verified that the updated vector $\mathbf{x}'$ thus defined is a feasible flow and the updated cost $z' = \mathbf{c}'\mathbf{x}' = \mathbf{c}\mathbf{x} + w\delta$, where w is the weight of the cycle C'. If C' is a negative cycle, $\mathbf{c}'\mathbf{x}' < \mathbf{c}\mathbf{x}$. Thus if the network $N'(\mathbf{x})$ has a negative cycle, we can obtain a better solution by updating the current solution $\mathbf{x}$ using the negative cycle. Equivalently, if the current solution $\mathbf{x}$ is optimal, there cannot be a negative directed cycle in $N'(\mathbf{x})$. The pleasant fact (as before) is that the converse is also true, giving the optimality criterion: $\mathbf{x}$ is optimal if and only if $N'(\mathbf{x})$ has no directed negative cycle. The presence of a negative cycle can be detected using the Floyd–Warshall algorithm. The method of obtaining an optimal solution from a given feasible solution by this iteration procedure is known as **algorithm cycle**. The proof is omitted here. For details refer to Papadimitriou and Steiglitz (1982).

Example 4.12

By using the algorithm cycle, obtain an optimal solution of the MCFP in the network shown below. Along each arc (i, j), there is a triple $(x_{ij}, u_{ij}; c_{ij})$ of three numbers with the usual meanings, where $0 \leqslant x_{ij} \leqslant u_{ij}$. The flow value is 4, which remains unchanged. The current cost is $z = 4 + 12 + 8 = 24$.

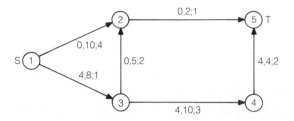

For the current flow **x**, the network $N(\mathbf{x})$ is as follows.

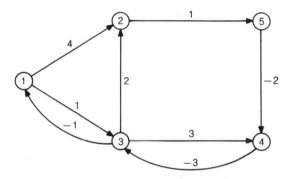

The Floyd–Warshall algorithm for $N'(\mathbf{x})$, at the fourth iteration, gives the following matrix which has a negative diagonal entry.

$$
\begin{bmatrix}
0 & 3 & 1 & 4 & 4 \\
M & 0 & M & M & 1 \\
-1 & 2 & 0 & 3 & 3 \\
-4 & -1 & -2 & 0 & 0 \\
-6 & -3 & -5 & -2 & -2
\end{bmatrix}
$$

So there is a negative cycle in $N'(\mathbf{x})$. This negative cycle can be located by using the corresponding path matrix. The negative cycle C' is $2 \to 5 \to 4 \to 3 \to 2$, with $w = 1 - 2 - 3 + 2 = -2$. The corresponding cycle C in $N(\mathbf{x})$ is $2 \to 5 \leftarrow 4 \leftarrow 3 \to 2$. The positive arcs in C are $(2, 5)$ and $(3, 2)$. The other two arcs in C are negative.

Now $\delta = \min\{2 - 0,\ 4,\ 4,\ 5 - 0\} = 2$. Thus the updated cost is $z' = 24 + w\delta = 24 - 4 = 20$. To get the updated flow, add two units to the current flow components along the positive arcs $(2, 5)$ and $(3, 2)$ in the cycle. At the same time, subtract two units from the current flow components along the negative arcs $(4, 5)$ and $(3, 4)$. The updated flow pattern is as shown below.

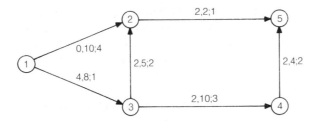

The current $N'(\mathbf{x})$ is as follows.

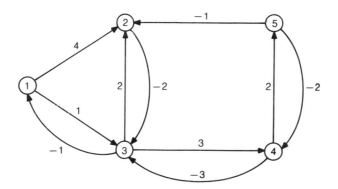

At the fifth iteration, the Floyd–Warshall algorithm for $N'(\mathbf{x})$ gives the following SD matrix for $N'(\mathbf{x})$, in which all the diagonal entries are 0:

$$\begin{bmatrix} 0 & 3 & 1 & 4 & 6 \\ -3 & 0 & -2 & 1 & 3 \\ -1 & 2 & 0 & 3 & 5 \\ -4 & 1 & -3 & 0 & 2 \\ -6 & 3 & -5 & -2 & 0 \end{bmatrix}$$

So the current $N'(\mathbf{x})$ has no negative cycles. Thus the current solution (the updated flow pattern displayed above) is an optimal solution of the MCFP with cost $4+4+2+6+4=20$.

Algorithm build-up

Suppose $\mathbf{x}$ is a known minimum cost flow with flow value r which is less than the maximum flow value v in the network. It is possible to obtain a minimum cost flow $\mathbf{x}'$ with a larger flow value $r' \leqslant v$ by updating $\mathbf{x}$. So if we take $\mathbf{x} = \mathbf{0}$ as the initial solution, we can ultimately obtain a minimum cost flow in the network with flow value equal to the maximum flow value. In this procedure, known as **algorithm build-up**, at each iteration the current flow value is increased by at least one unit. It may not be possible to reach v from r in one iteration. The flow value thus 'trickles up' by at least one unit at a time till it attains its maximum value while always maintaining the minimality of the cost. Our assumption all along is that the capacity vector is an integer vector. It is also assumed that the current flow value (corresponding to the current minimum cost flow $\mathbf{x}$) is a nonnegative integer r which is less than the maximum flow value v. Thus there exists a flow augmenting path in the network along which δ units of the flow can be shipped from the source to the sink. It is not at all necessary that the updated flow vector $\mathbf{x}'$, with the increased flow value $r' = r + \delta$, is a minimum cost flow. How do we update the current minimum cost flow $\mathbf{x}$ with flow value r to another feasible flow vector with flow value $r + \delta$ which is a minimum cost flow in the network? The following theorem gives us the answer.

Theorem 4.11

Let $\mathbf{x}$ be a minimum cost flow in a network with a flow value r which is less than the maximum flow value. Suppose the flow value is increased to $r + \delta$ by using a flow augmenting path which corresponds to a minimum weight (shortest) path in $N'(\mathbf{x})$ from source to sink. Then the updated flow is a minimum cost flow with flow value $r + \delta$.

Proof

Suppose the updated flow $\mathbf{x}'$ is not a minimum cost flow. Then the network $N'(\mathbf{x}')$ will have a negative cycle C' which should have at least one arc (q, p) with negative weight. So the flow augmenting path P which was used to increase the flow value from r to $r + \delta$ should have used the original arc (p, q) in the network to augment the flow. So P is of the form

$$1 - \cdots p \rightarrow q - \cdots - n$$

(an arc in P other than (p, q) could be forward or backward).

Let D be the part of the negative cycle C' obtained from C' by deleting the arc (q, p). Suppose $w(D)$ is the sum of the weights of the arcs in D. Then

$$w(D) + (-c_{pq}) < 0$$

which implies $w(D) < c_{pq}$.

Now consider the path P' in the network from 1 to n consisting of the following three consecutive subpaths: (a) the path from 1 to p as in P; (b) the path from p to q consisting of the arcs from D; and (c) the path from q to n as in P. Obviously P' is a flow augmenting path. Now

$$w(P) - w(P') = c_{pq} - w(D) > 0$$

which implies $w(P') < w(P)$. This contradicts the fact that P is a minimum weight flow augmenting path.

Example 4.13
In the network of Example 4.12, the updated flow **x** is a minimum cost flow with flow value 4. Use algorithm build-up and obtain a minimum cost flow with a higher flow value.

The network $N'(\mathbf{x})$ for the current flow is also as depicted in Example 4.12. A path of minimum weight in $N'(\mathbf{x})$ from source (vertex 1) to sink (vertex 5) is $1 \rightarrow 3 \rightarrow 4 \rightarrow 5$. The corresponding flow augmenting path in $N(\mathbf{x})$ is the same. Using this path we can send two more units from source to sink. The updated flow as displayed below is a minimum cost flow with flow value 6.

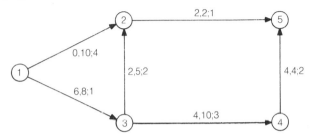

Transforming one flow problem into another

Theorem 4.12
A capacitated transshipment problem can be transformed into a transportation problem. In particular, every minimum cost flow problem with a preassigned permissible flow value can be transformed into a transportation problem.

Proof
Let $G = (V, E)$ be the capacitated network with n vertices and m arcs associated with a capacitated transshipment problem. The capacity vector is $\mathbf{u} = [u_{ij}]$ and the cost vector is $\mathbf{c} = [c_{ij}]$. We now construct a weighted bipartite graph $G' = (L, R, F)$ as follows.

Corresponding to each arc (i, j) in G, we define a supply vertex S_{ij} (as a left vertex in L) with supply equal to its capacity u_{ij}. Corresponding to each vertex i in G we define a demand vertex D_i (as a right vertex in R).

If i is a supply vertex in G, then the demand at D_i is the sum of the capacities of all the arcs directed from i minus the supply at i. If i is a demand vertex in G, then the demand at D_i is the sum of the capacities of all arcs directed from i plus the demand at i. If i is an intermediate vertex, then the demand at D_i is the sum of the capacities of all arcs directed from i.

From each S_{ij}, we draw two arcs – one to D_i and the other to D_j. The cost associated with the former arc is 0 and the cost associated with the latter is c_{ij}. Thus we have a balanced transportation problem with m supply vertices, n demand vertices and $2m$ arcs.

Suppose $x = [x_{ij}]$ is a feasible flow in the capacitated network G. Define a flow in the bipartite graph G' as follows. Let the flow $f_{ij,j}$ from S_{ij} to D_j be x_{ij} and let the flow $f_{ij,i}$ from S_{ij} to D_i be $u_{ij} - x_{ij}$. Then the flow vector f thus defined is a feasible flow in G' and both x and f have the same cost. Suppose $f = [f_{ij,k}]$ is a feasible flow in the bipartite graph G'. Define a flow vector x in G as $x_{ij} = f_{ij,j}$. Then x is a feasible flow in G with the same cost as that of f.

Theorem 4.13
A transportation problem can be converted into a minimum cost flow problem.

Proof
Suppose G is the bipartite graph associated with a balanced transportation problem in which the supply vertices are S_i $(i = 1, 2, \ldots, m)$ and the demand vertices are D_j $(j = 1, 2, \ldots, n)$. The cost associated with the arc (S_i, D_j) is c_{ij}. Let v be total supply, which is equal to total demand.

Now we construct a directed graph G' such that G is a subgraph of G' with two more vertices S and T. The supply at S and the demand at T are both equal to v. In G', we draw arcs from S to each S_i with capacity equal to the supply at S_i and cost equal to 0. We also draw arcs from each D_i to T with capacity equal to the demand at D_i and cost equal to 0. The capacity of the arc (S_i, D_j) in G' is v and the cost is c_{ij}. In the capacitated network G' the vertices S_i and D_j are all intermediate vertices such that G' has a unique source S and a unique sink T.

Suppose x is a feasible solution of the transportation problem and let x_{ij} be the component of x along the arc (S_i, D_j). Then we can define a flow vector x' in G' such that the components of x' along the arcs (S, S_i), (S_i, D_j) and (D_j, T) are respectively the supply at S_i, x_{ij} and the demand at D_j. Likewise, if x' is a feasible solution in G', we can define a flow vector x in G. It is easy to see that x is feasible if and only if x' is feasible, and that both x and x' have the same cost.

Corollary 4.14
The minimum cost flow problem and the transportation problem are equivalent.

Example 4.14
The capacitated transshipment problem (below, top) and the transportation problem (below, bottom) are equivalent. In the former, each arc is assigned a pair in which the first number is the flow and the second is the capacity. The cost vector is not given explicitly.

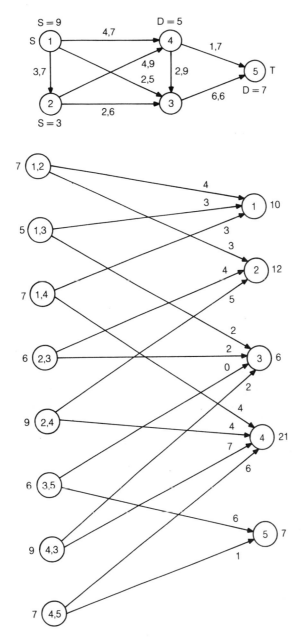

Theorem 4.15
The maximum flow problem can be converted into a capacitated transshipment problem.

Proof

Let $G = (V, E)$ be a capacitated network with $V = \{1, 2, \ldots, n\}$, where i is the source and n is the sink. Assume there is no arc from n to 1. Let $G' = (V, F)$, where $F = E \cup \{(n, 1)\}$. Let $c_{n1} = -1$ and $c_{ij} = 0$ for every other arc in G'.

We now assign a weight to the arc $e = (n, 1)$ so that G' becomes a capacitated network. This weight is a fixed positive number which could be taken as the sum of the capacities of all the arcs in the network. Let each vertex in G' be an intermediate vertex, including the vertices 1 and n. So we have a capacitated balanced transshipment problem on the directed network G'. A feasible solution in G with maximum flow value is an optimal solution in G' and vice versa.

Theorem 4.16

The feasibility problem in a capacitated transshipment problem can be transformed into a maximum flow problem.

Proof

Let G be the capacitated network associated with a balanced transshipment problem with total supply v. Construct a directed graph G' such that G is a subgraph of G' with two more vertices S and T. If i is a supply vertex in G, then draw arc (S, i) with capacity equal to the supply at i. If j is a demand vertex, draw arc (j, T) with capacity equal to the demand at j. Now G' is a capacitated network. An optimal solution of the maximum flow problem in G' with flow value equal to v will give a feasible solution of the transshipment problem. In fact we have the following feasibility criterion: the transshipment problem in G is feasible if and only if the maximum flow value in G' is v.

Theorem 4.17

The bipartite cardinality matching problem can be transformed into a minimum cost flow problem.

Proof

The bipartite graph is $G = (V, W, E)$. We construct a graph G' with two more vertices S and T, such that G is a subgraph of G'. Draw arcs from S to each vertex in V and from each vertex in W to T. The capacities of all arcs in G' are 1. Any feasible flow in G' with integer components is a matching in G. Conversely, any matching in G defines a feasible flow in G'. If (i, j) is a matched arc in G, the flows along (s, i), (i, j) and (j, t) are all 1. Exhaust all arcs in the matching like this. Let the flow in other arcs of G' be 0. Thus we have a feasible flow in G'. Thus every matching of cardinality v defines a feasible flow in G' with flow value v.

4.8 FLOW PROBLEMS AND COMBINATORICS

In this section we discuss some combinatorial implications of the max-flow min-cut theorem.

Theorem 4.18
The max-flow min-cut theorem implies Konig's theorem.

Proof
Let $G = (X, Y, E)$ be a bipartite graph where X is the set of left vertices and Y is the set of right vertices. Construct a directed graph $G' = (V', E')$ as follows. V' is the union of X, Y and a set of two new vertices – a source s and a sink t. Each edge in G now becomes an arc from a vertex in X to a vertex in Y. Construct artificial arcs from s to each vertex in X and artificial arcs from each vertex in Y to t. The capacity of each artificial arc is 1 and the capacity of every other arc is infinite.

Obviously any feasible flow in G' with flow value k corresponds to a matching of cardinality k in G, and vice versa. So a flow with maximum flow value in G' defines a matching with maximum cardinality in G.

Consider any covering W in G consisting of r vertices. The set W can be partitioned into two subsets $W(X)$ and $W(Y)$, where $W(X)$ is a set of left vertices and $W(Y)$ is a set of right vertices. Since W is a covering there is no arc in G' directed from a vertex in $X' = X - W(X)$ to a vertex in $Y' = Y - W(Y)$. (See Example 4.15 below.)

Let $S = \{s\} \cup W(Y) \cup X'$ and $T = \{t\} \cup W(X) \cup Y'$. Then (S, T) is an s–t cut in G' and the capacity of this cut is the cardinality of W. Thus any covering in G consisting of r vertices defines an s–t cut in G' with cut value equal to r.

On the other hand, consider any s–t cut (S, T) in G' with a finite cut value r. Since any arc from a vertex in X to a vertex in Y is of infinite capacity, every arc in the cut (S, T) is an artificial arc. So the cut has r arcs.

We now construct a set W of vertices as follows. If (s, x) is in (S, T), then x is in W. If (y, t) is in (S, T), then y is in W. Then W is a covering with r vertices. So there is a one-to-one correspondence between coverings in G and s–t cuts in G'. Thus any minimum s–t cut in G' defines a minimum covering in G. Hence the max-flow min-cut theorem implies Konig's theorem.

Example 4.15
The bipartite graph shown below

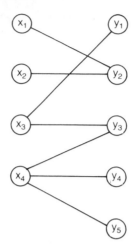

is expanded into the following digraph:

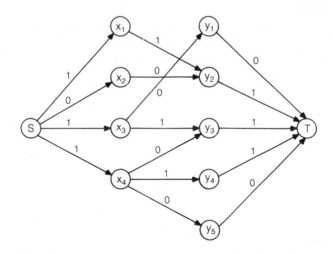

A matching in G is $M = \{(x_1, y_2), (x_3, y_3), (x_4, y_4)\}$. The corresponding feasible flow (with flow value 3) in G' can be shipped from s to t using the arcs in M. $W = \{x_3, x_4, y_2\}$ is a covering in G. Thus

$$X' = X - \{x_3, x_4\} = \{x_1, x_2\}$$
$$Y' = Y - \{y_2\} = \{y_1, y_3, y_4, y_5\}$$

(S, T) is a cut in G' where $S = \{s, x_1, x_2, y_2\}$ and $T = \{t, x_3, x_4, y_1, y_3, y_4, y_5\}$. Thus $(S, T) = \{(s, x_3), (s, x_4), (y_2, t)\}$. $|M| = |W| = |(S, T)| = $ flow value 3. So M is a maximum matching in G, W is a minimum covering G, (S, T) is a minimum s–t cut in G' and the maximum flow value in G' is 3.

Vertex-capacitated networks

Let $G = (V, E)$ be a digraph, where $V = \{1, 2, \ldots, n\}$. In addition to the arc capacity vector $\mathbf{u} = [u_{ij}]$, we now introduce a nonnegative vertex capacity vector $\mathbf{w} = [w_i]$. Vertex 1 is the unique source and vertex n is the unique sink; we shall denote these two vertices by s and t, respectively.

A nonnegative flow vector $\mathbf{x} = [x_{ij}]$ is an s–t **feasible flow** if, in addition to the usual conservation conditions and the arc capacity constraints, it also satisfies the following $n - 2$ vertex constraints: the outflow at every intermediate vertex cannot exceed the capacity of that vertex.

Observe that if the vertex capacities are all infinite, then the vertex constraints are always satisfied and any feasible flow in the usual sense is a flow in this more general context.

A **generalized s–t cut** is a set X of arcs and vertices such that any directed path from s to t in G will use at least one member of the set X. The capacity of a generalized cut is the sum of the capacities of its elements. In the digraph shown below, the set $\{(1, 2), (2, 3), (3, 6), 5\}$ is a generalized s–t cut.

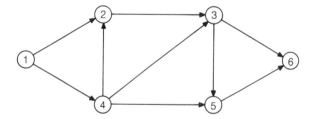

If all the vertex capacities are infinite, then a minimum s–t generalized cut X is a set which consists of arcs only. Let S be the set of all vertices that can be reached from the source s by directed paths that do not use any arc from the set X, and let $T = V - X$. Then (S, T) is an s–t cut in the conventional sense and every arc in (S, T) is an arc in X. Thus the minimum cut capacity in the usual sense is the same as the minimum cut capacity in this generalized sense when the vertex capacities are all infinite.

Theorem 4.19 (**Generalized max-flow min-cut theorem**)
The maximum value of a generalized s–t flow is equal to the capacity of a minimum s–t generalized cut.

Proof
The s–t network G is expanded into a larger s–t network G' by replacing each intermediate vertex i by two intermediate vertices i' and i'' connected by the arc (i', i'') of capacity w_i. Each arc (i, j) of G is replaced by an arc (i'', j'). Let $s' = s'' = s$ and $t' = t'' = t$. (See Example 4.16 below.) Thus G' is an s–t directed graph which is arc-capacitated but not vertex-capacitated. Any flow entering i' must pass through i'' and all flow leaving i'' must come from i'. So there is

a one-to-one correspondence between generalized feasible flows in G and feasible flows in G'. Thus the max-flow min-cut theorem in G' implies the max-flow min-cut theorem in G.

Example 4.16
Obtain a generalized maximum flow and the corresponding generalized cut in digraph G shown below. The capacities of arcs and vertices are displayed in the network.

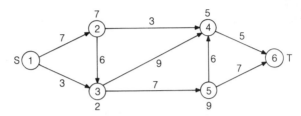

The expanded network G' is as follows:

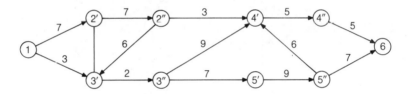

The maximum flow value in G' is 5. A minimum cut in G' is $\{(2'', 4'), (3', 3'')\}$, with cut value 5. The arc $(2'', 4')$ corresponds to the arc $(2, 4)$ in G. The arc $(3', 3'')$ corresponds to the vertex 3 in G. Thus a generalized minimum cut in G is $\{(2, 4), 3\}$. The generalized maximum flow is as shown below:

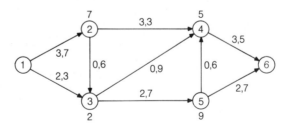

Menger's theorems

Let $G = (V, E)$ be a digraph with a source s and a sink t. A set W of intermediate vertices in G is an s–t **vertex-separating set** in G if there is no directed path from s to t in the subgraph obtained from G after deleting from V all the vertices belonging to W. The digraph G is said to be **k-vertex-connected** from s to t if there exists an s–t vertex-separating set of cardinality k and no set with less than k intermediate vertices is an s–t vertex-separating set. In other words, G is k-vertex-connected from s to t if and only if, for any set W of $k-1$ intermediate vertices, there should be a directed path from s to t which would not use any vertices from the set W. A set F of arcs in G is an **s–t arc-separating set** in G if there is no directed path from s to t in the subgraph obtained from G after deleting from E all the arcs belonging to F. The digraph G is said to be **k-arc-connected** from s to t if there exists an s–t arc-separating set of cardinality k and no set with less than k arcs is an s–t arc-separating set. A set C of arcs is an **s–t cut** in G if C is the set of all arcs in G directed from a vertex in a set S (which contains s and does not contain t) to a vertex in $T = V - S$. Any s–t cut is an s–t arc-separating set. Two directed paths from s to t in the digraph are said to be **vertex-disjoint** if they have no vertices in common other than s and t, and **arc-disjoint** if they have no arcs in common. The definitions of these concepts in the case of undirected graphs are analogous.

We are now ready to state and prove four different versions of Menger's theorem involving graph connectivity.

Theorem 4.20 (Vertex form of Menger's theorem for directed graphs)
If s and t are two distinct vertices in a digraph with no arc from s to t, then the maximum number of vertex-disjoint directed paths from s to t equals the minimum number of vertices whose deletion destroys all directed paths from s to t in the digraph.

Proof
If there are k vertex-disjoint paths from s to t, then the digraph is certainly k-vertex-connected from s to t. The crux of the theorem is in the converse part.

Assign a capacity of 1 to each vertex and infinite capacity to each arc. Since there is no arc from s to t, the set of all intermediate vertices is a generalized s–t cut. So the capacity of any generalized minimum s–t cut is necessarily finite. Since the graph is k-connected, the cardinality of this minimum cut (which is actually an s–t vertex-separating set) is k. So the maximum flow value is k.

Since we can send only one unit of the flow along any directed path from s to t, there should be k vertex-disjoint directed paths from s to t. Thus the largest number of vertex-disjoint directed paths from s to t equals the smallest number of vertices which must be deleted to disconnect all directed paths from s to t. This completes the proof.

Theorem 4.21 (Arc form of Menger's theorem for directed graphs)
If s and t are distinct vertices in a digraph, then the maximum number of arc-disjoint directed paths from s to t equals the minimum number of arcs whose deletion destroys all directed paths from s to t in the digraph.

Proof

If G is a digraph with a source s and a sink t and if there are k arc-disjoint directed paths from s to t in G, then G is obviously k-arc-connected from s to t. We now prove the converse using the max-flow min-cut theorem.

Assign a capacity of 1 to each arc in the digraph $G = (V, E)$. Suppose K is a minimum s–t arc-separating set. Let $G' = (V, E - K)$. Let X be the set of all vertices that are reachable from the source in G'. Then $(X, V - X)$ is an s–t cut in G. Every arc in K is an arc directed from some vertex in X to some vertex in $V - X$, and vice versa. In other words, the s–t arc-separating set K is actually an s–t cut. So if a digraph is k-arc-connected from s to t and if the capacity of each arc is 1, then there exists a minimum s–t cut with k arcs, and by the max-flow min-cut theorem there is a flow with flow value k.

We now prove by induction that there are k arc-disjoint paths from s to t in G. Since the capacities are all integers, we can assume that the components of any feasible flow are integers.

Let I be the set of all positive integers n such that if there is a flow with integer components with flow value n, then there are n arc-disjoint paths from s to t. Obviously $1 \in I$. Assume that $(k - 1) \in I$. Suppose the flow value is k. There is a path P from s to t along which a flow of one unit can be sent. The capacity of each arc in P is 1. Delete all the arcs belonging to P from the digraph. The flow value in the resulting digraph is $(k - 1) \in S$. So these $k - 1$ arc-disjoint paths and the path P together constitute a set of k arc-disjoint paths. So $k \in I$. This completes the induction argument.

If G is an undirected graph with a source s and a sink t, we can replace an edge joining two vertices i and j by two arcs (i, j) and (j, i), each with capacity 1. By applying the previous theorems to the digraph thus obtained, we obtain the following undirected versions of Menger's theorems.

Theorem 4.22 (Vertex form of Menger's theorem for undirected graphs)
If s and t are distinct vertices in an undirected graph with no edge between s and t, the maximum number of vertex-disjoint paths between s and t equals the minimum number of vertices whose deletion destroys all paths between s and t in the graph.

Theorem 4.23 (Edge form of Menger's theorem for undirected graphs)
If s and t are distinct vertices in an undirected graph, then the maximum number of edge-disjoint paths between s and t equals the minimum number of edges whose deletion destroys all paths between s and t in the graph.

Theorem 4.24
Menger's theorem implies Konig's theorem.

Proof
Let $G = (X, Y, E)$ be a bipartite graph. Construct the digraph G' as in Theorem 4.18. Apply Theorem 4.22 to G'.

Theorem 4.25
Menger's theorem implies the max-flow min-cut theorem.

Proof
Assume that the capacity of each arc is a nonnegative integer. Construct a multidigraph $G' = (V, E')$ corresponding to the capacitated s–t network $G = (V, E)$ as follows. If the capacity of the arc (i, j) in G is c_{ij}, then draw c_{ij} arcs from i to j in G'. The value of a maximum flow in G is equal to the total number p of arc-disjoint paths from s to t in G', and the value of a minimum cut in G is equal to the minimum number q of arcs in an s–t arc-separating set in G'. By the arc form of Menger's theorem, the numbers p and q are equal. Thus the maximum value of a flow from s to t in G is equal to the minimum value of an s–t cut in G.

Lower bound constraints and circulations

Consider an upper-bounded (capacitated) network $G = (V, E)$ in which each arc (i, j) has a nonnegative lower bound constraint $d(i, j)$ in addition to the upper bound constraint $u(i, j)$, where $d(i, j) \leqslant u(i, j)$ for each arc (i, j). Thus both **d** and **u** are nonnegative functions on E such that $\mathbf{d} \leqslant \mathbf{u}$. A flow vector **f** in G is a **feasible flow** in G if $d(i, j) \leqslant f(i, j) \leqslant u(i, j)$ for each arc (i, j) in the network. In an arbitrary network equipped with both upper bound and lower bound constraints, a feasible flow need not exist. Consider, for example, the network $G = (V, E)$ with $V = \{1, 2, 3\}$ and $E = \{(1, 2), (2, 3)\}$, in which vertex 1 is the source and vertex 3 is the sink with $u(1, 2) < d(2, 3)$.

A feasible flow in a network equipped with both lower bound and upper bound constraints is called a **circulation** in the network if the netflow at each vertex is 0. So there is no supply vertex or demand vertex in this case. Every vertex is considered as an intermediate vertex. In particular, there is no vertex designated as a source or as a sink.

A circulation problem in $G = (V, E)$ equipped with a lower bound constraint **d** and upper bound constraint **u** can be converted into a flow problem in an upper-bounded network $G' = (V', E')$ with a source s and a sink t by introducing a new vertex s (the source) and a new vertex t (the sink) and by drawing arcs from s to each vertex in V and from each vertex in V to t. V' is the union of V and the set of these two vertices. E' is the union of E (the set of original arcs) and the set of new arcs thus constructed.

In the enlarged s–t network G', we define an upper bound constraint u' and a flow $\mathbf{f}'$ as follows. If the arc (i, j) is an original arc, its upper bound constraint $u'(i, j)$ is $u(i, j) - d(i, j)$ and its flow component $f'(i, j)$ is $f(i, j) - d(i, j)$. The upper bound constraint $u'(s, i)$ of the new arc (s, i) is by definition of the sum of the lower bounds of all the arcs directed to i in G. The flow component $f'(s, i)$ is defined to be the same as $u'(s, i)$. Likewise, the upper bound constraint of the new arc (j, t) is the sum of the lower bounds of all the arcs directed from j in G, and the flow component $f'(j, t)$ is $u'(j, t)$.

The following figure shows a network G in which the notation x, y, z along an arc (i, j) means $d(i, j) = x, f(i, j) = y$ and $u(i, j) = z$.

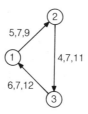

Here we have the enlarged s–t network showing the current flow pattern:

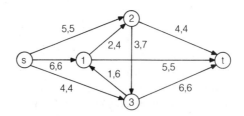

Theorem 4.26 (Hoffman's circulation theorem)
In a network $G = (V, E)$ equipped with an upper bound constraint function **u** and a lower bound constraint function **d** in which every vertex is an intermediate vertex, a necessary and sufficient condition for the existence of a circulation is that $d(V - W, W) \leqslant u(W, V - W)$ for every set W of vertices in G.

Proof
If **f** is a circulation in G and if W is a set of vertices in G, then

$$f(W, V) = f(V, W)$$

and

$$f(W, V - W) = f(V, W) - f(W, W) = f(V - W, W)$$

Consequently,

$$d(V - W, W) \leqslant f(V - W, W) = f(W, V - W) \leqslant u(W, V - W)$$

Thus

$$d(V - W, W) \leqslant u(W, V - W)$$

for every set W of vertices. So this condition is necessary.

Let us write $\bar{W} = V - W$. Suppose $d(\bar{W}, W) \leqslant u(W, \bar{W})$ for every subset W of V. A circulation **f** in G will exist if and only if there exists a flow **f**′ in the enlarged network $G' = (V', E')$ such that the flow value $f'(s, V)$ is equal to

$d(V, V)$. If W' is any set of vertices in G', let $\bar{W}'$ be the set $V' - W'$. The max-flow min-cut theorem in G' implies $f'(s, V) \leqslant u'(W', \bar{W}')$. Thus a circulation f will exist in G if and only if $d(V, V) \leqslant u'(W', \bar{W}')$ for any cut $(W', \bar{W}')$ in G'. So it is enough if we show that this inequality is valid in G'.

Suppose $(W', \bar{W}')$ is a cut in G'. Let $W = W' - s$. Then $\bar{W} = V - W = \bar{W}' - t$. Now

$$u'(W', \bar{W}') = u'(W \cup s, \bar{W} \cup t)$$
$$= u'(W, \bar{W}) + u'(s, \bar{W}) + u'(W, t) \text{ in } G'$$
$$= u(W, \bar{W}) + d(V, \bar{W}) + d(W, V) - d(W, \bar{W}) \text{ in } G$$
$$= u(W, \bar{W}) + d(W, \bar{W}) + d(\bar{W}, W) + d(W, V) - d(W, \bar{W})$$
$$= u(W, \bar{W}) + d(W, \bar{W}) + d(W, V)$$
$$= [u(W, \bar{W}) - d(\bar{W}, W)] + d(V, V) \geqslant d(V, V) \text{ by hypothesis}$$

Thus the condition is sufficient.

Theorem 4.27

The max-flow min-cut theorem and Hoffman's circulation theorem are equivalent.

Proof

According to Theorem 4.26, the max-flow min-cut theorem implies the circulation theorem. Now given a flow in an upper-bounded s–t network $G' = (V, E')$, we construct an enlarged network $G = (V, E)$ where E is obtained by adjoining one more arc (t, s) to the set E'. (Without loss of generality we assume that there is no arc (t, s) in G'.) The upper bound for (t, s) is $+\infty$ and the upper bound for every other arc is the same in both G and G'. The lower bound for the arc (t, s) is the minimum cut value q in the network G', and the lower bound for every other arc in G is 0. Then any circulation in G corresponds to a flow in G' with flow value p greater than or equal to q. Thus the circulation theorem implies the max-flow min-cut theorem if the network G has a circulation. A sufficient condition for the existence of a circulation is $u(W, V - W) - d(V - W, W) \geqslant 0$ for any set W of vertices in G. So all that remains to be established is that this inequality holds for any W.

There are four cases to be examined:

(i) $s, t \in W$. In this case $u(W, V - W)$ is nonnegative and $d(V - W, W)$ is zero.

(ii) $s \in W, t \notin W$. In this case

$$u(W, V - W) - d(V - W, W) = u(W, V - W) - q \geqslant 0$$

since q is a minimum cut value.

(iii) $s \notin W, t \in W$. In this case $u(W, V - W)$ is $+\infty$.

(iv) $s, t \notin W$. In this case $u(W, V - W)$ is nonnegative and $d(V - W, W)$ is zero.

This completes the proof.

Next we establish two generalizations of Hall's marriage theorem in the context of bipartite graphs.

Theorem 4.28 (Weighted bipartite matching theorem)

Let $G = (X, Y, E)$ be a bipartite graph, where $X = \{x(i): i \in I = \{1, 2, \dots, m\}\}$ and $Y = \{y(j): j \in J = \{1, 2, \dots, n\}\}$. The edge joining $x(i)$ and $y(j)$ is denoted by (i, j). Suppose a and b are functions from X and Y, respectively, to the set of nonnegative integers. Let $a(Z) = \sum_{x(i) \in Z} a(x(i))$, where $Z \subset X$, and $b(Z') = \sum_{y(j) \in Z'} b(y(j))$, where $Z' \subset Y$. Assume that $a(X) = b(Y)$. Under these conditions the following properties are equivalent.

1. There exists a function w from E to the set of nonnegative integers such that the (edge weights) $w(i, j)$ of the edges (i, j) satisfy the following two equations:

   (i) $\displaystyle\sum_{j, (i, j) \in E} w(i, j) = a(x(i))$ for each i

   (ii) $\displaystyle\sum_{i, (i, j) \in E} w(i, j) = b(y(j))$ for each j.

2. $a(S) \leqslant b(f(S))$ for all $S \subset X$, where $f(S)$ is the set of vertices in Y which are adjacent to at least one vertex in S.

Proof

Suppose property 1 holds. Let S be any subset of X. The set of edges which are adjacent to at least one vertex in S is a subset of the set of edges which are adjacent to at least one vertex in $f(S)$ and therefore $a(S) \leqslant b(f(S))$. So property 1 implies property 2.

To prove the reverse implication, we construct an enlarged bipartite graph $G' = (X', Y', E')$ as follows. Replace each vertex $x(i)$ with $a(i)$ vertices and each vertex $y(j)$ with $b(j)$ vertices. Let $X(i)$ be the set of vertices which replace $x(i)$ and $Y(j)$ be the set of vertices which replace $y(j)$. Join every vertex in $X(i)$ and every vertex in $Y(j)$ if and only if there is an edge in G joining $x(i)$ and $y(j)$. Let E' be the set of edges thus constructed by taking into account each edge in E. Finally, let $X' = \cup X(i)$ and $Y' = \cup Y(j)$.

Let $T \subset X'$. Define $T' = \cup \{X(i): T \cap X(i) \neq \varnothing\}$. Even though $|T| \leqslant |T'|$, $|f(T)| = f(T')$. Let $S = \{x(i) \in X : T \cap X(i) \neq \varnothing\}$. Then $a(S) = |T'|$ and $b(f(S)) = |f(T')|$. Now property 2 implies $a(S) \leqslant b(f(S))$ and therefore $|T'| \leqslant |f(T')|$. But $|T| \leqslant |T'|$ and $|f(T)| = |f(T')|$. Thus $|T| \leqslant |f(T)|$ for any subset T of X'. So by Hall's marriage theorem there is a complete matching from X' to Y' in G'.

Let the number of matched edges between the vertices in the set $X(i)$ and the set $Y(j)$ be $w(i, j)$. Then $\sum_j w(i, j) = a(i)$ for each i since the matching is complete from X to Y, and $\sum_i w(i, j) \leqslant b(j)$ for each j since the number of matched edges incident to $b(j)$ vertices in $Y(j)$ is at most $b(j)$. Suppose there exists a vertex $y(j)$ such that $\sum_i w(i, j) < b(j)$. Then

$$a(X) = \sum_i a(i) = \sum_i \sum_j w(i, j)$$

$$= \sum_j \sum_i w(i, j) < \sum_j b(j) = b(Y)$$

But $a(X) = b(Y)$ by hypothesis. Thus $\sum_i w(i, j) = b(j)$ for each j. Hence property 2 implies property 1.

Notice that this theorem reduces to Hall's marriage theorem if $a(x(i)) = 1$ for each i and $b(y(j)) = 1$ for each j.

Example 4.17
Consider the bipartite graph $G = (X, Y, E)$, where $X = \{x(1), x(2)\}$ with $a(1) = 3$ and $a(2) = 4$, and $Y = \{y(1), y(2), y(3)\}$ with $b(1) = 2$, $b(2) = 3$ and $b(3) = 2$. The edges are $(1, 1), (1, 2), (2, 1), (2, 2)$ and $(2, 3)$. It can be easily verified that property 2 is satisfied. The edge weights can be taken as $w(1, 1) = 1$, $w(1, 2) = 2$, $w(2, 1) = 1$, $w(2, 2) = 1$ and $w(2, 3) = 2$.

Theorem 4.29 (Constrained weighted bipartite matching theorem)
Let $G = (X, Y, E)$ be a bipartite graph satisfying the same conditions as in Theorem 4.28. Assume, furthermore, that G is a complete bipartite graph and that each edge (i, j) has a nonnegative integer weight $c(i, j)$. Under these conditions the following properties are equivalent:

3. There exists a function w from E to the set of nonnegative integers such that the edge weights $w(i, j)$ satisfy the following three conditions:

   (i) $w(i, j) \leqslant c(i, j)$ for each edge (i, j) in E

   (ii) $\sum_j w(i, j) = w(x(i), Y) = a(x(i))$ for each i in I

   (iii) $\sum_i w(i, j) = w(X, y(j)) = b(y(j))$ for each j in J

4. $a(S) \leqslant b(T) + c(S, \bar{T})$ for all $S \subset X, T \subset Y$.

Proof
Suppose property 3 holds. Let $S \subset X$ and $T \subset Y$. Then

$$a(S) = w(S, Y) = w(S, T) + w(S, \bar{T}) \leqslant w(X, T) + c(S, \bar{T}) = b(T) + c(S, \bar{T})$$

Thus property 3 implies property 4.

To establish the reverse implication, we enlarge the complete bipartite graph G into another (not complete) bipartite graph $G^* = (X^*, Y^*, E^*)$ as follows. On each edge (i, j) joining $x(i)$ and $y(j)$, we insert two vertices $y(i, j)$ and $x(i, j)$ of degree 2 such that the edge (i, j) is divided into three edges as $\{x(i), y(i, j)\}, \{y(i, j), x(i, j)\}$ and $\{x(i, j), y(j)\}$. Let X' be the set of all $x(i, j)$

vertices thus constructed on all the edges and let Y' be the set of all $y(i, j)$ vertices. Let $X^* = X \cup X', Y^* = Y \cup Y'$ and E^* be the set of new edges thus constructed.

Since $|X| = m$ and $|Y| = n$, there will be $3mn$ edges in E^*. Also there will be m vertices of degree n and mn vertices of degree 2 in X^*. Likewise there will be n vertices of degree m and mn vertices of degree 2 in Y^*.

Introduce two weight functions a^* and b^* on X^* and Y^* by defining

$$a^*(x(i)) = a(x(i))$$
$$b^*(y(j)) = b(y(j))$$
$$a^*(x(i, j)) = b^*(y(i, j)) = c(i, j)$$

so that G^* becomes a weighted bipartite graph. Let $S \subset X$ and $f(S)$ be the set of all vertices in Y which are adjacent to at least one vertex in S. Then property 4 implies

$$a(S) \leqslant b(f(S)) + c(S, \overline{f(S)})$$

So

$$a(S) + c(S, f(S)) \leqslant b(f(S)) + c(S, Y)$$

The left-hand side of this inequality is the maximum value of $a^*(S)$ and the right-hand side is the minimum value of $b^*(f(S))$. Thus property 4 implies property 2 of Theorem 4.28 as far as the bipartite graph G^* is concerned. So there exist edge weights on the edges in G^* satisfying the conditions stated as in property 1 of Theorem 4.28.

Let the weights on the edges $\{x(i), y(i, j)\}$, $\{y(i, j), x(i, j)\}$ and $\{x(i, j), y(j)\}$ be $w(i, j)$, $u(i, j)$ and $w'(i, j)$, respectively. Then

$$a^*(x(i, j)) = w'(i, j) + u(i, j)$$
$$b^*(y(i, j)) = w(i, j) + u(i, j)$$

But, by definition,

$$a^*(x(i, j)) = b^*(y(i, j)) = c(i, j)$$

Therefore $w(i, j) = w'(i, j)$ for each (i, j). In other words, the edge joining $x(i)$ and $y(j)$ has weight $w(i, j)$ which is less than or equal to $c(i, j)$. Moreover, $w(x(i), Y) = a^*(x(i))$ and $w(X, y(j)) = b^*(y(j))$ in G^* according to Theorem 4.29. But $a^*(x(i)) = a(x(i))$ and $b^*(y(j)) = b(y(j))$. Thus property 3 is satisfied.

Theorem 4.30
The constrained weighted bipartite matching theorem implies Hoffman's circulation theorem.

Proof
Let $G = (V, E)$ be a digraph, where $V = \{v_i : i = 1, 2, \ldots, n\}$. If there is an arc from v_i to v_j, then there are two given nonnegative integers $d(i, j)$ and $u(i, j)$ such

that $d(i, j) \leqslant u(i, j)$. Here $d(i, j)$ is the lower bound and $u(i, j)$ is the upper bound of the arc. If there is no such arc, then we assume that $d(i, j) = u(i, j) = 0$. So without loss of generality we assume that there is an arc from every vertex to every other vertex in G.

Construct a complete bipartite graph $G' = (X, Y, E')$ in which $X = \{x_1, x_2, \ldots, x_n\}$ and $Y = \{y_1, y_2, \ldots, y_n\}$. On each edge joining x_i and y_j, we define a constraint $c(i, j)$ which is equal to $u(i, j) - d(i, j)$ if $i \neq j$. We also define $c(i, i) = +\infty$ for each i.

Let D be the sum of all lower bounds and U be the sum of all upper bounds. Let M be chosen such that $M > D + U$. Define

$$a(x(i)) = M - d(v(i), V)$$
$$b(y(j)) = M - d(V, y(j))$$

Then G' is a weighted complete bipartite graph with a constraint defined on each edge. Let $S \subset X$ and $T \subset Y$. Then

$$a(S) = |S|M - d(S, Y)$$
$$b(T) = |T|M - d(X, T)$$

In particular, $a(X) = b(Y)$. (Notice that in our notation $a(S)$ we consider S as a subset of X. But when we write $d(S, Y)$ we consider S as a subset of V and Y as the set V. This lack of precision does not cause any problem in our discussion.)

Let S be any subset of V. To establish the Hoffman circulation theorem we have to show that the condition $d(\bar{S}, S) \leqslant u(S, \bar{S})$ is equivalent to the existence of a circulation in G.

Now $d(\bar{S}, S) \leqslant u(S, \bar{S})$ implies $d(V, S) - d(S, S) \leqslant u(S, \bar{S})$. So

$$d(V, S) - [d(S, V) - d(S, \bar{S})] \leqslant u(S, \bar{S})$$

Hence

$$d(V, S) - d(S, V) \leqslant c(S, \bar{S})$$

Thus

$$[|S|M - b(S)] - [|S|M - a(S)] \leqslant c(S, \bar{S})$$

which implies the inequality

$$a(S) \leqslant b(S) + c(S, \bar{S}) \qquad \text{for all } S \subset X \qquad (*)$$

Now suppose S is an arbitrary subset of X and T an arbitrary subset of Y. Our aim is to establish the inequality

$$a(S) \leqslant b(T) + c(S, \bar{T}) \qquad (**)$$

for the complete bipartite graph G' so that Theorem 4.29 can be applied.

Recall that S and T can also be considered as subsets of V. So (**) holds when $S = T$ because of the inequality (*). We have to examine cases when S and T are not equal.

If S is a proper subset of T, $|S|$ is less than $|T|$ and therefore, by making M sufficiently large, we can make (**) hold good.

In the remaining cases $c(S, \bar{T}) = +\infty$. Thus (**) holds for any $S \subset X$ and $T \subset Y$. So according to Theorem 4.29, there exist edge weights $f'(i, j)$ such that

$$f'(i, j) \leqslant c(i, j), \quad f'(i, Y) = a(x(i))$$
$$f'(X, j) = b(y(j))$$

Now $0 \leqslant f'(i, j) \leqslant u(i, j) - d(i, j)$ implies

$$d(i, j) \leqslant f'(i, j) + d(i, j) \leqslant u(i, j)$$

Let $f(v_i, v_j) = f(i, j) = f'(i, j) + d(i, j)$. Then the outflow from v_i is

$$f(v_i, V) = f(i, Y) = f'(i, Y) + d(i, Y) = a(x(i)) + d(i, Y) = M$$

Likewise, the inflow to v_j is

$$f(V, v_j) = f(X, j) = f'(X, j) + d(X, j) = b(y(j)) + d(X, j) = M$$

Thus $d(\bar{S}, S) \leqslant c(S, \bar{S})$ implies there exists a circulation in G.

On the other hand, if there is a circulation in G, it is easy to show that the inequality is satisfied as established at the beginning of Theorem 4.26.

Thus Theorem 4.29 implies the Hoffman circulation theorem.

Theorem 4.31
Hall's marriage theorem implies Hoffman's circulation theorem.

Proof
Hall's marriage theorem implies Theorem 4.28 which implies Theorem 4.29. But Theorem 4.29 implies the circulation theorem as proved in Theorem 4.30.

On combining the results established in this section and in section 2.5 we have the following equivalence theorem.

Theorem 4.32
Dilworth's theorem, Hall's marriage theorem, Hoffman's circulation theorem, Konig's theorem, the Konig–Egervary theorem, the max-flow min-cut theorem and Menger's theorem are equivalent.

Figure 4.10 illustrates the various implications.

4.9 EXERCISES

1. Consider the network shown below with flow vector $\mathbf{x}^T = [x_{12} \ x_{13} \ x_{23}$ $x_{34} \ x_{45} \ x_{46} \ x_{51} \ x_{62} \ x_{65}]$, cost vector $\mathbf{c} = [6 \ 12 \ 0 \ 6 \ 3 \ 9 \ 0 \ 3 \ 9]$ and supply–demand vector $[0 \ 2 \ 4 \ -14 \ 0 \ 8]$.

   (i) Take $[2 \ 4 \ 0 \ 0 \ 14 \ 6 \ 0 \ 8]^T$ as an initial feasible tree solution and solve the uncapacitated transshipment problem.

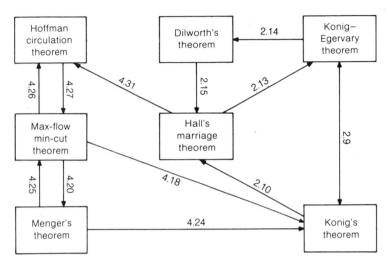

Fig. 4.10 Implications and equivalences of theorems in combinatorics and graph theory. Numbers refer to theorems in the text.

(ii) Suppose the flow vector **x** has the upper bound constraint **u** where $\mathbf{u}^T = [4\ 8\ 2\ 2\ 14\ 14\ 6\ 2\ 14]$. Solve the corresponding capacitated transshipment problem by taking $[2\ 2\ 2\ 0\ 4\ 10\ 4\ 2\ 0]^T$ as an initial feasible tree solution.

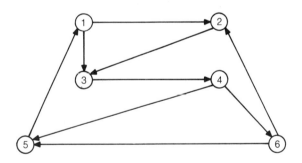

2. Consider the network shown below with flow vector $\mathbf{x}^T = [x_{13}\ x_{21}\ x_{32}$ $x_{51}\ x_{52}\ x_{61}\ x_{64}\ x_{67}\ x_{74}\ x_{75}]$, cost vector $\mathbf{c} = [4\ 4\ 4\ 4\ 20\ 44\ 80$ $20\ 20\ 24]$ and supply–demand vector $[0\ 6\ 9\ 15\ -6\ -9\ -15]$.

(i) Take $[9\ 9\ 0\ 0\ 15\ 0\ 9\ 0\ 6\ 9]^T$ as an initial feasible tree solution and solve the uncapacitated transshipment problem.

(ii) Suppose the flow vector **x** has the upper bound constraint **u** where $\mathbf{u}^T = [15\ 3\ 9\ 12\ 9\ 18\ 9\ 15\ 27\ 3]$. Solve the corresponding capacitated transshipment problem by taking $[9\ 3\ 0\ 0\ 9\ 6\ 3\ 0\ 12\ 3]^T$ as an initial feasible tree solution.

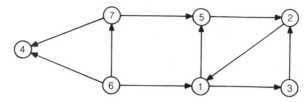

3. Let G be the network shown below with flow $\mathbf{x}^T = [x_{12}\ x_{13}\ x_{15}\ x_{23}\ x_{42}\ x_{43}\ x_{53}\ x_{54}\ x_{56}\ x_{62}\ x_{64}]$. The capacity vector is $[2\ 10\ 10\ 6\ 8\ 9\ 9\ 10\ 6\ 7\ 8]$, the cost vector is $\mathbf{c} = [1\ 2\ 3\ 4\ 6\ 5\ 4\ 3\ 2\ 1\ 2]$ and the supply–demand vector is $[-9\ 4\ 17\ 1\ -5\ -8]$. Test whether $[0\ 0\ 9\ 6\ 4\ 3\ 8\ 0\ 6\ 6\ 8]$ is an optimal solution. If it is not optimal, perform one iteration. By how much did the cost go down?

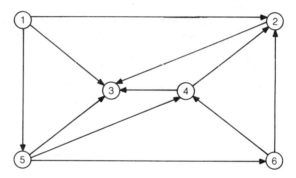

4. For the network shown below, suppose the supply–demand vector is $[0\ 0\ 12\ 20\ 16\ -18\ -30]$ and the flow $\mathbf{x}^T$ is $[x_{13}\ x_{14}\ x_{15}\ x_{21}\ x_{23}\ x_{24}\ x_{25}\ x_{54}\ x_{61}\ x_{62}\ x_{63}\ x_{67}\ x_{72}\ x_{75}]$. Test whether $[0\ 20\ 0\ 2\ 12\ 0\ 0\ 0\ 18\ 0\ 0\ 0\ 14\ 16]$ is an optimal solution if the capacity of each arc is 20 and $\mathbf{c}$ is $[106\ 36\ 58\ 16\ 120\ 56\ 74\ 10\ 88\ 76\ 196\ 28\ 46\ 118]$.

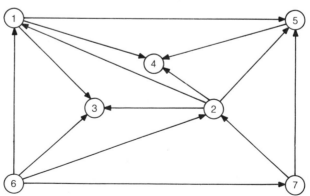

5. The supply–demand vector in the network shown below is

$$[-6 \ -8 \ -10 \ 0 \ 0 \ 0 \ 10 \ 8 \ 6]$$

Construct two artificial arcs (2, 5) and (5, 8) to obtain a feasible flow in the enlarged network. Use this feasible flow to obtain an FTS in the network.

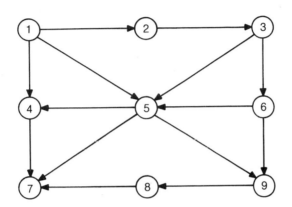

6. Consider the capacitated network shown below, with supply–demand vector $[-10 \ -8 \ -6 \ 9 \ 0 \ 15]$ and with capacity vector $[9 \ 5 \ 12 \ 7 \ 10 \ 9 \ 8 \ 10 \ 6]$. Construct an additional arc (5, 6) with unlimited capacity and obtain an FTS for the enlarged problem. Show that the given problem is feasible.

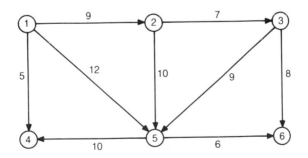

7. In the network shown below, the flow vector is $x^T = [x_{18} \ x_{23} \ x_{27} \ x_{36} \ x_{42} \ x_{46} \ x_{51} \ x_{54} \ x_{61}]$. The supply–demand vector is $[-2 \ 2 \ 4 \ -6 \ -3 \ 5 \ 7 \ -7]$. Show that this uncapacitated problem can be converted into two subproblems.

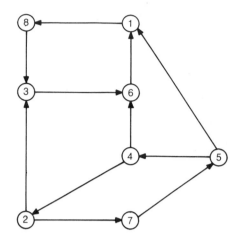

8. In the network shown below, the flow vector is $\mathbf{x}^T = [x_{23}\ x_{25}\ x_{26}$ $x_{31}\ x_{34}\ x_{41}\ x_{46}\ x_{51}\ x_{56}]$, with capacity $\mathbf{u}^T = [2\ 9\ 7\ 9\ 8\ 9\ 8\ 2\ 7]$, where $\mathbf{b}^T = [13\ -12\ -3\ -6\ 5\ 3]$. Show that this capacitated problem can be converted into two subproblems.

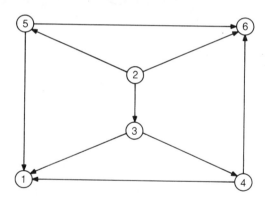

9. In the network $G = (V, E)$, where $V = \{1, 2, 3, 4, 5, 6\}$, with flow vector $\mathbf{x}^T = [x_{21}\ x_{26}\ x_{32}\ x_{43}\ x_{45}\ x_{61}\ x_{63}]$, the supply–demand vector is $[10\ -2\ 0\ -5\ 0\ -3]$. The component of the cost vector is 1 along each arc. The capacity of each arc is 10, except for that of the arc $(3, 2)$ whose capacity is only 5.

   (i) Solve this problem by inspection.
   (ii) Show that $W = \{1, 2, 6\}$ is an autonomous set.
   (iii) Obtain the enlarged network G' by constructing artificial arcs $(2, 5)$, $(6, 5)$, $(5, 1)$ and $(5, 3)$, and solve the enlarged problem.
   (iv) Show that one choice of T^* consists of arcs $(4, 3)$, $(3, 2)$, $(2, 1)$, $(6, 1)$ and the artificial arc $(6, 5)$ along which there is no flow.

(v) Decompose the original problem into two subproblems, solve each subproblem individually and merge them to obtain an optimal solution for the original problem.

10. Show that it is possible for an infeasible capacitated transshipment problem to have an autonomous set by constructing an example.

11. Obtain the maximum flow value and the minimum cut in the network shown below, using the Edmonds–Karp method.

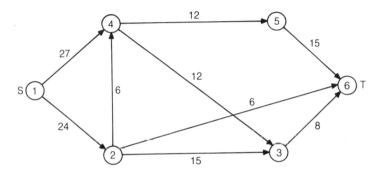

12. Obtain the maximum flow value and the minimum cut in the network shown below, using the Edmonds–Karp method.

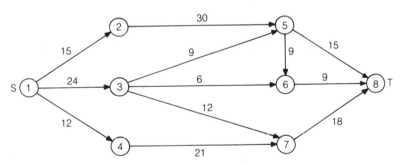

13. Obtain the maximum flow value and the minimum cut in the network shown below, using the Edmonds–Karp method.

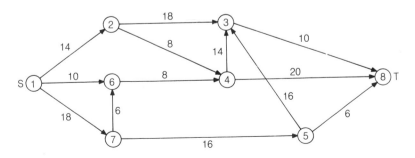

14. Obtain the maximum flow value and the minimum cut in the network shown below, using the Edmonds–Karp method.

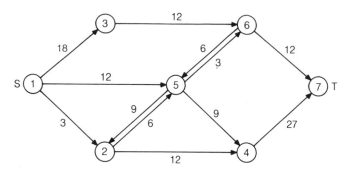

15. Obtain the maximum flow value and the minimum cut in the network shown below, using the MPM algorithm.

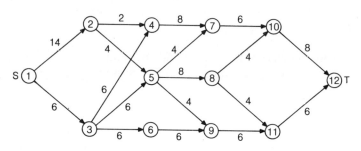

16. Obtain the maximum flow value and the minimum cut in the network shown below, using the MPM algorithm.

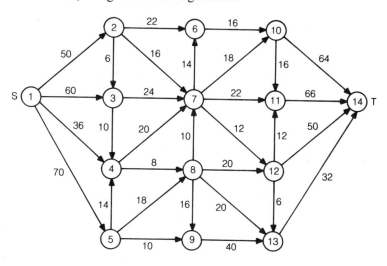

17. Obtain the maximum flow value and the minimum cut in the network shown below, using the MPM algorithm.

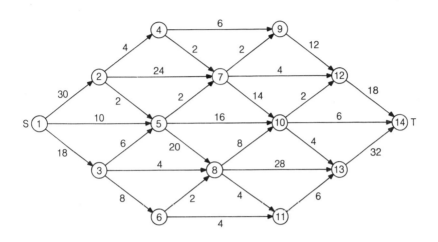

18. Obtain the maximum flow value and the minimum cut in the network shown below, using the MPM algorithm.

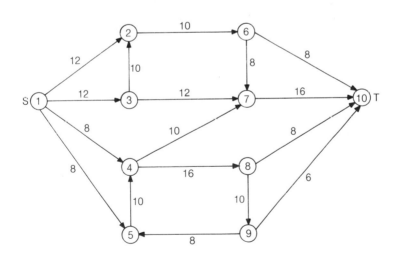

19. Obtain the maximum flow value and the minimum cut in the network shown below, using the MPM algorithm.

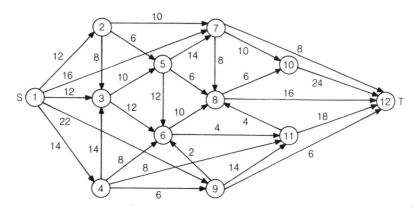

20. Consider the network in Exercise 15 as an undirected network. Solve the corresponding maximum flow problem as an undirected planar problem, using the procedure outlined in section 4.5.
21. Consider the network in Exercise 16 as an undirected network. Solve the corresponding maximum flow problem as an undirected planar problem, using the procedure outlined in section 4.5.
22. Consider the network in Exercise 17 as an undirected network. Solve the corresponding maximum flow problem as an undirected planar problem, using the procedure outlined in section 4.5.
23. Obtain the Gomory–Hu tree corresponding to the multiterminal flow problem related to the network shown below. (This problem is from Hu, 1982.)

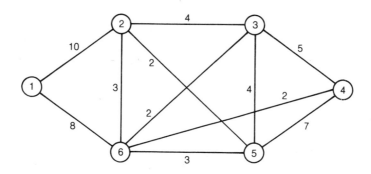

24. Obtain the Gomory–Hu tree corresponding to the multiterminal flow problem related to the network shown below.

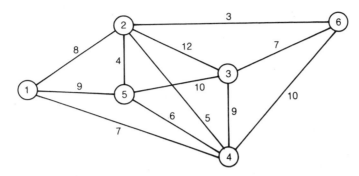

25. Obtain the Gomory–Hu tree corresponding to the multiterminal flow problem related to the network shown below.

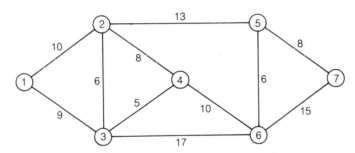

26. Convert Exercise 20 of section 2.7 (a transportation problem) into a minimum cost flow problem and then obtain an optimal solution by using the algorithm cycle.

27. Convert Exercise 20 of section 2.7 (a transportation problem) into a minimum cost flow problem and then obtain an optimal solution by using algorithm build-up.

5

Matchings in graphs

5.1 CARDINALITY MATCHING PROBLEMS

A matching in an undirected graph $G = (V, E)$ is a set M of edges such that no two edges in M have a vertex in common. The problem of finding a matching in a given graph with as many edges as possible is known as the **cardinality matching problem**. If each edge of the graph is assigned a weight, the **weighted matching problem** is the problem of finding a matching such that the sum of the weights of all the edges in the matching is as large as possible. Matching problems involving bipartite graphs are known as assignment problems, and we have already discussed algorithms to solve problems of this kind in previous chapters (section 2.6 for the weighted case and section 4.8 for the nonweighted case) as a part of the analysis of flow problems in uncapacitated and capacitated networks. Bipartite matching problems are easy to solve because they can be modeled as network flow problems.

In this chapter our goal is to study procedures for solving both the cardinality matching problem and the weighted matching problem in an arbitrary undirected graph. In both types of problem, the problem is easier to investigate in the bipartite case. So we look at bipartite graphs first and then make appropriate generalizations to the nonbipartite case. Needless to say, it is the existence of odd cycles in nonbipartite graphs that makes the matching problem in these graphs more difficult to solve.

An edge in a matching in a graph is called a **matched edge**, and an edge which is not in the matching is a **free edge**. If u and v are two vertices, each is a **mate** of the other if they are joined by a matched edge. A vertex is an **exposed vertex** if it is not incident to a matched edge. A vertex which is not exposed is a **matched vertex**.

A simple path P between two vertices is an **alternating path** with respect to M if its edges are alternately free and matched. An alternating path between two vertices u and v with respect to M is an **augmenting path** with respect to M if both u and v are exposed. An augmenting path with k matched edges will have $k+1$ free edges and $2k+2$ vertices. An augmenting path with respect to M can be used to increase the cardinality of M, and this is the content of the following theorem.

Theorem 5.1
If P is an augmenting path with respect to a matching M in a graph G, the symmetric difference M' between M and P is also a matching in G and $|M'| = |M| + 1$.

Proof
Let e and f be any two edges in M'. We prove that e and f have no vertex in common.

The set M' is the disjoint union of $M - P$ and $P - M$. If both e and f are in $M - P$, they both are in M and so they cannot have a vertex in common. If both are in $P - M$, they cannot have a vertex in common since P is an alternating path.

Suppose $e \in (M - P)$ and $f \in (P - M)$. Assume P is a path between u and v. The only unexposed vertices in P are u (the first vertex in P) and v (the last vertex in P). So if e and f have a vertex in common, that vertex must be either u or v. In either case the stipulation that the first and last vertices of an augmenting path have to be exposed is violated. Thus M' is indeed a matching.

Suppose P has k matched edges and $k+1$ free edges. Then

$$|M - P| = |M| - k$$
$$|P - M| = 2k + 1 - k = k + 1$$

Thus

$$|M'| = |M| - k + k + 1 = |M| + 1$$

This completes the proof.

So if there is an augmenting path with respect to a given matching, then that matching is not a maximum cardinality matching. As in the case of flow networks, the converse also is true. This is our next theorem.

Theorem 5.2
A matching is a maximum cardinality matching if and only if there is no augmenting path with respect to the matching.

Proof
It is enough if we prove that if M is not a maximum cardinality matching, then there exists an augmenting path with respect to M.

Suppose M' is a maximum cardinality matching. Then $|M| < |M'|$. Let $G' = (V, E')$, where $E' = M \oplus M'$. The degree in G' of any vertex v cannot be more than 2 since two edges belonging to the same matching cannot have a vertex in common. If the degree is 2, of the two edges incident at a vertex, one should belong to M and the other to M'. So every connected component of G' is either an alternating path or a cycle.

In any connected component which is a cycle the number of edges belonging to M will be the same as the number of edges belonging to M'. So if all the connected components of G' are cycles both M and M' will have the same cardinality, which is indeed not the case. So at least one component of G' is an alternating path P in which both the first edge and the last edge belong to M', since M' has more edges than M. But this path is an augmenting path for M since its first vertex and the last vertex are both exposed as far as M is concerned.

So if M is not a maximum cardinality matching, there exists an augmenting path with respect to M.

Cardinality matching in bipartite graphs

We can increase the number of edges in a nonoptimal cardinality matching if we can locate an augmenting path with respect to that matching. By making use of augmenting paths and increasing the cardinality one at a time, eventually (after a finite number of iterations) we obtain a matching for which there is no augmenting path. At this stage we have an optimal cardinality matching. What we need, therefore, is a systematic procedure for locating augmenting paths. In the case of bipartite graphs there is a straightforward method for accomplishing this.

Suppose M is a matching in a bipartite graph $G = (X, Y, E)$, where X is the set of left vertices and Y is the set of right vertices. If M is not a maximum cardinality matching, then there exists an augmenting path P between an exposed vertex in X and an exposed vertex in Y. Our aim is to start from an exposed vertex $x \in X$ and try to construct an augmenting path to an exposed vertex $y \in Y$.

If we can obtain an alternating path P between an exposed left vertex x_i and a matched left vertex x_j which is joined by an edge e to an exposed right vertex y_k, then an augmenting path between x_i and y_k is readily available by adjoining e to P. This observation leads us to define a digraph $G' = (X, F)$ on the set of left vertices as follows. Whenever we have a free edge between a left vertex x_i and a right vertex y_k and a matched edge between y_k and x_j, draw an arc from x_i to x_j.

Any directed path in the digraph thus constructed from an exposed vertex u to a vertex v which is joined by a free edge to a right vertex w will give an augmenting path between u and w. The matched left vertex v serves as an 'accomplice' for the exposed left vertex u because in the updated matching v will help two exposed vertices become matched vertices without losing its own matched status! If there are two exposed left vertices, then there is no directed path in G' from one to the other. If there are no exposed left vertices, then the current matching is optimal. If there are exposed left vertices and at the same time there is no vertex in G' which could serve as an accomplice, then the matching is again optimal.

Thus given a matching M in the bipartite graph which has at least one left exposed vertex and at least one right exposed vertex, the first job is to see whether there exists an edge between two such vertices. If there is an edge of this type, then the two exposed vertices are matched by this edge. If we continue like this we will reach either of the following stages: there is no exposed left vertex or no exposed right vertex; or there is at least one exposed left vertex and at least one exposed right vertex with no edge joining them. In the former case, we have obtained an optimal matching. In the latter case, we construct the corresponding digraph G' in which we carry out a breadth-first search starting from an exposed left vertex. As soon we reach a left vertex in G' which is joined by a free edge to a right exposed vertex in G, we stop and update the matching by increasing its cardinality by 1. Here is an illustrative example.

Example 5.1

In the bipartite graph shown below, the edges of the current matching M are drawn as double lines. The free edges are drawn as single lines. The left vertices are x_i $(i = 1, 2, \ldots, 6)$ and the right vertices are y_i $(i = 1, 2, \ldots, 5)$. Test whether M is an optimal cardinality matching. If it is not optimal, obtain as many augmenting paths as possible.

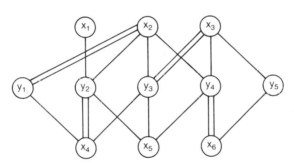

In the current matching the left vertices x_1, x_5 and the right vertex y_5, as shown above, are exposed vertices. The vertex x_3 is a matched left vertex which is joined by a free edge to the right exposed vertex y_5. The vertex x_3 is a candidate for becoming an accomplice. Similarly, x_6 is also a candidate for this status. So if we can obtain an alternating path between x_1 and x_3 or x_6 in G (or between x_5 and x_3 or x_6), then we can increase the cardinality of M. The corresponding digraph G' is as shown below.

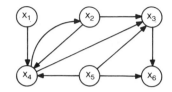

We start a BFS search from x_1 in G'. The BFS tree is as follows:

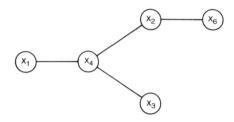

This search gives two directed paths in G', one from x_1 to the accomplice x_3, and the other from x_1 to x_6. The BFS tree emanating from x_5 is as follows:

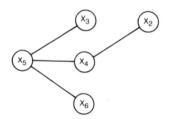

There are three directed paths in G', one from x_5 to x_6, another from x_5 to x_3, and another from x_5 to x_2. Thus, there are five augmenting paths for this bipartite graph:

(a) $x_1 - y_2 - x_4 - y_1 - x_2 - y_4 - x_6 - y_5$
(b) $x_1 - y_2 - x_4 - y_3 - x_3 - y_5$
(c) $x_5 - y_4 - x_6 - y_5$
(d) $x_5 - y_3 - x_3 - y_5$
(e) $x_5 - y_2 - x_4 - y_1 - x_2 - y_4$

Cardinality matching in nonbipartite graphs

As in the case of bipartite graphs, corresponding to a given matching M in a nonbipartite graph $G = (V, E)$ we can construct a directed graph $G' = (V, F)$ in which we draw an arc from v_i to v_j whenever v_j has a mate v_k which is joined to v_i by a (free) edge. An augmenting path in G between v and w will always correspond to a directed path from v to a matched vertex u which is joined to w by a free edge.

But the converse is not generally true in the case of a nonbipartite graph. Suppose 1 is an exposed vertex and 5 is a matched vertex which is joined to an exposed vertex v by an edge in G. Assume there is a directed path in G' from 1 to 5 as shown below:

$$1 \rightarrow 2 \rightarrow 3 \rightarrow 4 \rightarrow 5$$

Since the graph is not bipartite the corresponding undirected path in G may have repeated vertices. Suppose this path is as follows, in which the matched edges are represented by double lines:

$$1—5=2—8=3—7=4—2=5—v$$

If we make each matched edge a free edge and each free edge a matched edge in this path with a view to augmenting the cardinality, then we see that 2 will be matched with both 4 and 8. So this path (with repeated vertices) cannot be an augmenting path.

This path with repeated vertices can be symbolically represented as in the diagram below, pointing out the presence of an odd cycle (with as many matched edges as possible) which is causing the trouble.

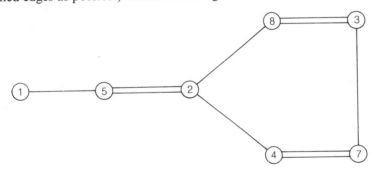

An odd cycle in itself does not cause this problem. It is easy to construct examples to confirm this. The problem arises only when out of the $2k+1$ edges in an odd cycle there are k matched edges. Such densely matched odd cycles are called **blossoms** by Edmonds (1965a). More precisely, any cycle C with $2k+1$ edges, of which k are matched (with respect to a matching M) is called a blossom in G with respect to M if there is an alternating path P (the **stem** of the blossom) from an initial exposed vertex u (the **root** of the blossom) which enters the cycle C through a vertex v (the **base** of the blossom) which is incident to two free edges in the cycle. Every vertex in a blossom is a matched vertex. The base is the only vertex v whose mate is not in the blossom. If the current matching does not create any blossoms, then the graph for all practical purposes can be treated as a bipartite graph even though there could be odd cycles which are not blossoms. Thus we have to obtain a procedure which will first locate blossoms whenever they occur and obtain augmenting paths even in their presence.

Locating a blossom

In our search for an augmenting path from an exposed vertex u, suppose we move along the stem of a blossom B with root at u. When we reach the base v of B there are two free edges incident at v of which only one can be chosen for inclusion in the alternating path. We enter the blossom along one edge and return to the base by leaving the blossom through the other edge. Observe that if we enter the blossom through a vertex which is not the base, then there is only one choice for the next edge in our alternating path.

We consider a blossom with a base at vertex 1 as shown below and a BFS search which starts from its base:

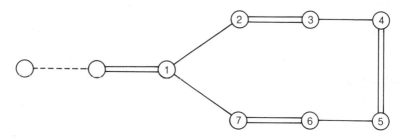

One alternating path is 1—2=3—4=5. Another alternating path is 1—7=6—5=4. We see that the edge (4, 5) appears in both of them. This is an indication that a blossom exists. So if the BFS search gives two alternating paths from a common vertex and if both paths have an edge in common as the $(k+1)$th edge, then a blossom with $2k+1$ edges exists, the base of which can be obtained by simultaneously backtracking from the common edge along the two paths till we reach a common vertex which is the base. Thus the presence of a blossom is detected and its base is identified.

Alternating trees

A matching, as we proved earlier, is a maximum cardinality matching if and only if there is no augmenting path with respect to that matching. The maximum cardinality matching algorithm is based on this fact. In the algorithm, the search is for an augmenting path from an exposed vertex by constructing a tree rooted at an exposed vertex. By our construction, every path in the tree between the root and a vertex is an alternating path in which the first edge is free. So the tree is called an **alternating tree**. If the last edge in the (unique) path from the root to a vertex v is a matched edge, then v is called an **outer vertex**. By definition, the root is an outer vertex. A vertex which is not outer is an **inner vertex**. The degree (in the tree) of an inner vertex is either 2 or 1.

Suppose $\{x, y\}$ is an edge of the graph to be examined for possible inclusion in the current alternating tree T while we are at vertex x in T. Vertices in T are **labeled vertices** and edges in T are **marked edges**. If x is an inner vertex and if $\{x, y\}$ is not a matched edge, then $\{x, y\}$ is not to be included. If $\{x, y\}$ is a matched edge, then it is included in T and y becomes an outer vertex. If x is an outer vertex, then there are four possibilities regarding the eligibility of the free edge $\{x, y\}$ for possible inclusion in the tree.

(a) If y is not in T (y is not labeled) and is exposed, then an augmenting path has been found from the root to y in which the last edge is $\{x, y\}$ which is included in T.
(b) If y is not in T and if y is matched with a mate z (which is also not a vertex in T), then both $\{x, y\}$ and $\{y, z\}$ are included in the tree with y as an inner vertex and z as an outer vertex.
(c) If y is an inner vertex in T, then the inclusion of the edge $\{x, y\}$ will create an even cycle. Hence the edge $\{x, y\}$ is not to be included in T.
(d) If y is an outer vertex in T, then the edge $\{x, y\}$ creates a blossom.

Thus the procedure to construct an alternating tree is as follows.

Step 1. Start from an exposed vertex u which is the root of the tree. Label u as an outer vertex. All other vertices are unlabeled and all edges are unexamined.

Step 2. Choose an unexamined edge $\{x, y\}$ where x is an outer vertex. (If there is no such edge, go to step 4.)

(a) If y is not a labeled vertex, then it is labeled as an inner vertex and the edge $\{x, y\}$ is marked for inclusion in the tree. If y is an exposed vertex, then an augmenting path from u to y has been found and we stop. If y has a mate z, then the matched edge $\{y, z\}$ is marked for inclusion in the tree. The vertex z is labeled as an outer vertex. Repeat step 2.
(b) If y is an inner vertex, then mark $\{x, y\}$ as an edge not in the tree. Repeat step 2.
(c) If y is an outer vertex, then go to step 3.

Step 3. Since both x and y are outer vertices and the unmarked edge now becomes a marked edge, the edge $\{x, y\}$ is free. This situation creates a blossom. We stop.

Step 4. At this stage, it is not possible to mark any more edges for the tree. All vertices of degree 1 (other than the root) are outer vertices. An alternating tree of this type is called a **Hungarian tree**. There is no augmenting path from u. We stop.

So the procedure in the construction of an alternating tree from an exposed vertex u has three possible outcomes:

(a) We may obtain an augmenting path starting from u with respect to M in the graph, using which we can obtain a matching M' with $|M'| = |M| + 1$.
(b) We may end up with a Hungarian tree in which every alternating path from u ends up in a matched vertex, indicating that there are no augmenting paths (with respect to M) from u.
(c) We may reach an outer vertex, indicating the existence of a blossom.

If the search from an exposed vertex u results in a Hungarian tree, then there is no augmenting path from u (with respect to the current matching) in the graph. If the current matching M is not optimal, then there is a matching M' which has more elements than M. The surprising fact is that there is no augmenting path from u with respect to the matching M', which is the content of the next theorem. Consequently, if all the exposed vertices with respect to the current matching M produce Hungarian trees, then M is an optimal cardinality matching.

Theorem 5.3

If u is an exposed vertex with respect to a nonoptimal matching M in an undirected graph, and if there are no augmenting paths from u with respect to M, then there is no augmenting path from u with respect to M', where M' is a matching with more edges than M.

Proof

Since M is nonoptimal and since there is no augmenting path from u, there is an augmenting path P (with respect to M) between two exposed vertices v and w. Let $M' = M \oplus P$. Suppose an augmenting path Q (with respect to M') exists, starting from the vertex u and ending in the vertex t.

Case 1: $P \cap Q = \varnothing$. The first edge f in Q is not in M'. So it is not in $M - P$. By assumption, f is not in P either. So f is not in M. Similarly, the last edge in Q is not in M. The second edge in Q is in M' but not in P. So it is in M. Thus Q is an alternating path with respect to M in which the first and last edges are free with respect to M. In other words, Q is an augmenting path with respect to M from u. This is a contradiction.

Case 2: $P \cap Q \neq \varnothing$. In this case the two paths have a vertex in common. Let u_j be the first vertex (counting from v) on P that is common with the path q (see the diagram below). Then u_j divides P into two subpaths (one starting from v and the other starting from w) such that in one of these subpaths the last edge is in M. Let P' be the subpath in which the last edge is in M. Let Q' be the subpath of Q from u to u_j. Let R be the path obtained by joining P' and Q'. Every matching edge in P' is in M and the first edge in P' is not a matched edge. The first edge in Q' is not in M. Every matched edge in Q' is in M' but not in P. So it is in M. Thus R is an augmenting path with respect to M starting from u. This is a contradiction.

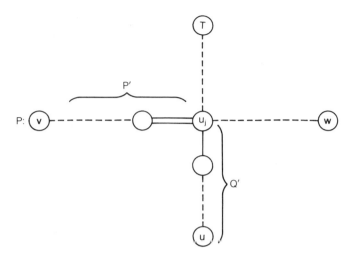

Corollary 5.4
If all the exposed vertices with respect to the current matching M produce Hungarian trees, then M is an optimal cardinality matching.

Shrinking a blossom

Let M be a matching in $G = (V, E)$. Suppose in the search for an augmenting path a blossom is located. Let B be the set of all vertices in this blossom. We now shrink all the blossom vertices into a single vertex v_B. Let $V/B = (V - B) \cup \{v_B\}$. We now define a graph $G/B = (V/B, E/B)$ as follows:

1. Any edge in G joining two vertices in $V - B$ continues as an edge in G/B.
2. Any edge joining a vertex u in $V - B$ and a vertex v in B is replaced by an edge joining u and v_B.
3. The matched edges in G continue as matched edges in G/B. We thus shrink the blossom into a single vertex and the graph G into G/B. The matching in G/B is denoted by M/B.

Theorem 5.5
A matching M in a graph G is a maximum cardinality matching if and only if M/B is a maximum cardinality matching in G/B, where B is any blossom with respect to M in G.

Proof
It is enough if we show that G has an augmenting path with respect to M if and only if G/B has an augmenting path with respect to M/B.

Let P be an augmenting path in G/B with respect to M/B. Let v_B be the condensed vertex in G/B. If P does not pass through v_B, then P is an augmenting path in G.

Now consider an augmenting path P in G/B which passes through v_B. Either v_B is a matched vertex or an exposed vertex.

If v_B is a matched vertex, then the path P defines the following three paths in G (see the diagram below) in sequence when the blossom is expanded:

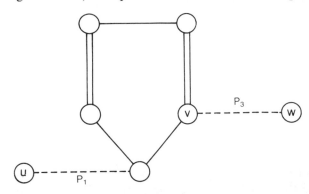

(i) P_1: an alternating path from an exposed (with respect to M) vertex u to the base of the blossom in which the last edge is matched.
(ii) P_2: an alternating path in the blossom from the base to a vertex v in which the last edge is a matched edge.
(iii) P_3: an alternating path from v to a vertex w in which both the first and last edges are free. Then $P_1 \cup P_2 \cup P_3$ is an augmenting path (with respect to M) in G between u and w.

If v_B is an exposed vertex, then P is an augmenting path in G/B starting from v_B. So P is an augmenting path (see the diagram below) in G from a vertex v in the blossom. Let Q be a path in the blossom (when the blossom is expanded) from its base to v in which the last edge is a matched edge. Then the union of Q and P is an augmenting path in G from the base of the blossom to w.

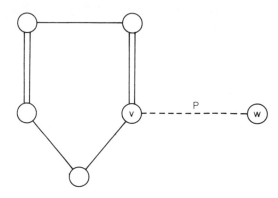

Thus any augmenting path in G/B defines an augmenting path in G.

Conversely, let us suppose P is an augmenting path in G. We have to show that there is an augmenting path in G/B. There are several cases to be examined:

(a) If P does not pass through v_B, then P is an augmenting path in G/B. If P enters the blossom at its base and does not go through any other vertex of the blossom, then P defines an augmenting path in G/B with v_B as an intermediate vertex. If P starts from the base of the blossom, then P defines an augmenting path in G/B with v_B as the starting vertex.

(b) Let P enter the blossom at its base b and leave the blossom at a vertex v. Then P consists (see the diagram below) of an alternating path P_1 from an exposed vertex u to the base b, then an alternating path P_2 (in the blossom) from b to v in which the last edge is a matched edge, and then an alternating path P_3 from v to an exposed vertex w. These three subpaths become the coalesced path $Q = (P_1, P_3)$ joined together by the condensed vertex v_B. Obviously Q is an augmenting path in G/B.

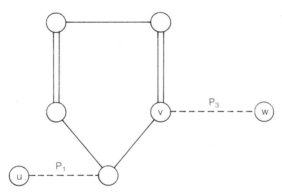

(c) Let P enter the blossom (see the diagram below) at a vertex v which is not the base of the blossom. Then P has to enter with a free edge. Suppose P leaves the blossom after traversing a path (in the blossom) in which the

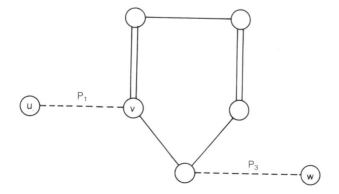

last edge is a free edge. The only way this can happen is when P leaves the blossom through its base. The path P consists of three disjoint alternating subpaths: the subpath P_1 from u to v, the subpath P_2 (in the blossom from v to the base) in which the first edge is a matched edge, and the subpath P_3 from the base to the exposed vertex w in which the first edge is a matched edge. Again the coalesced path (P_1, P_3) joined by v_B is an augmenting path in G/B.

(d) Let P (which starts from u) enter the blossom at a vertex v which is not the base and leave the blossom at a vertex v' after traversing a path (in the blossom) in which the last edge is a matched edge (see the diagram below). The first edge in P when it leaves the blossom is the free edge joining v' and v''. Let P_3 be the subpath of P from v'' to the exposed vertex w. Then the path P is the union of the following four subpaths in sequence: the subpath P_1 from the exposed vertex to the blossom vertex v; the subpath P_2 (in the blossom) from v to v'; the edge $\{v', v''\}$; and the subpath P_3. Notice that P does not pass through the base of the blossom. Since the blossom is discovered in a search for an augmenting path from the starting vertex u, there must be an alternating path Q from u to the base in which the last edge is a matched edge. If the paths Q and P_3 are disjoint, then the union of Q and P_3, joined by the condensed vertex v_B, is an augmenting path in G/B.

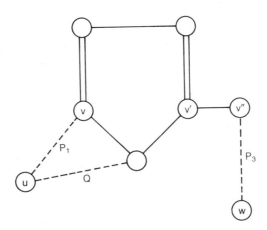

(e) Suppose the paths Q and P_3 of case (d) above are not disjoint. So there exists a vertex x in P_3 such that Q is the union of two paths Q' (from u to x) and Q'' (from x to the base) meeting at x (see the diagram below). The vertex x can be chosen such that Q and the subpath P'_3 of P_3 joining x and w have no vertex in common other than x. Then, under the assumption that P_1 and Q'' are disjoint, the union of P_1 and Q'' joined by v_B becomes an alternating path in G/B. The union of this alternating path and P'_3, joined by x, is an augmenting path in G/B.

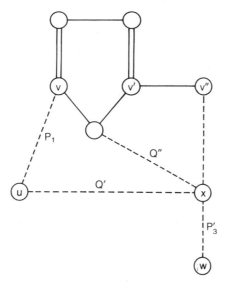

(f) As in (e) above, assume that Q and P_3 intersect. Now we relax the restriction that P_1 and Q'' are disjoint. Let y be a common vertex between P_1 and Q'' such that they have no vertex in common in their respective subpaths towards the blossom other than y. Either y is in P_3 or y is not in P_3. If y is in P_3 we can construct an augmenting path in G/B from u to v_B (along P_1), then from v_B to y along Q in the reverse direction, and finally from y to w along P_3. Otherwise we construct an augmenting path in G/B from u to y along P_1, then from y to v_B along Q, and then from v_B to w along P_3.

Thus all possibilities are exhausted, establishing the fact that the existence of an augmenting path in G implies the existence of an augmenting path in G/B.

Blossom within a blossom

If we obtain a blossom while starting from an exposed vertex v with respect to M in $G = (V, E)$, then we can obtain a condensed graph $G_1 = (V_1, E_1)$ and a matching M_1 by shrinking the blossom. If the number of exposed vertices in G_1 (with respect to M_1) is less than 2, then the matching M_1 is a maximum cardinality matching in G_1, which implies that M is a maximum cardinality matching in G. Otherwise, we start from an exposed vertex and proceed as before. At some stage, we end up with a condensed graph $G_k = (V_k, E_k)$ and a matching M_k which is a maximum cardinality matching in G_k. Recall G_k is obtained by shrinking a blossom B_{k-1} in G_{k-1}. We now replace the condensed blossom vertex in G_k by its original odd cycle and change the matching in this odd cycle into another maximum matching compatible with the matching in G_k. Then we have a maximum cardinality matching in G_{k-1}. We continue this process till we return to $G = (V, E)$.

At this stage we summarize our conclusions, which will lead to a procedure for obtaining a maximum cardinality matching in an undirected graph.

1. Let M be the current matching. If there are less than two exposed vertices, then the matching M is optimal.
2. If there are two or more exposed vertices, then start an alternating tree from an exposed vertex. If the alternating tree construction produces an augmenting path, then use this path to obtain the matching M' which has one more edge than M.
3. If the alternating tree starting from an exposed vertex with respect to the current matching is a Hungarian tree, then there is no augmenting path from that vertex with respect to the current matching and any subsequent matching involving more edges. In this case, we start an alternating tree from another exposed vertex.
4. If in the process of constructing an alternating tree from an exposed vertex we obtain a blossom B, then we shrink the blossom into a single vertex and construct the condensed graph G/B. Any maximum cardinality matching in G/B corresponds to a maximum cardinality matching in G, and vice versa. By opening the blossom, the maximum cardinality matching in G/B can be converted into a maximum cardinality matching in G.

The procedure for finding a maximum cardinality matching involves the seven steps listed below. A formal statement of the algorithm can be found in Lawler (1976), Papadimitriou and Steiglitz (1982) or Gondran and Minoux (1984). During the past few years, several modifications of Edmonds's blossom-shrinking procedure have been developed.

Step 1. The graph under consideration is G_0, with a matching M_0. All exposed vertices are unexamined. Let $i = 0$. Stop if there are less than two unexamined exposed vertices. Otherwise go to step 2.

Step 2. Start building an alternating tree rooted at an unexamined exposed vertex v. If an augmenting path is obtained, then go to step 3. If a blossom is obtained, then go to step 4. If a Hungarian tree is obtained, then go to step 5. If G_0 has only one unexamined exposed vertex, go to step 7.

Step 3. Use the augmenting path to increase the cardinality of M_i by 1. The root v is now a matched vertex. Go to step 6.

Step 4 (blossom shrinking). Let $i = i + 1$. Let B_i be the blossom. Shrink B_i into a condensed vertex b_i. The condensed graph is G_i, with matching M_i. Return to step 2.

Step 5 (Hungarian tree). The vertex v now becomes examined. Go to step 6.

Step 6 (blossom expansion). Let $G_1, G_2, \ldots, G_r$ be the graphs generated with condensed vertices $b_1, b_2, \ldots, b_r$ and matchings $M_1, M_2, \ldots, M_r$ in step 4. Let $M_r^* = M_r$. Open the blossom B_j to obtain the matching M_{j-1}^* from M_j^*, where $j = r, r - 1, \ldots, 1$. Go to step 2.

Step 7. The matching M_0^* is a maximum cardinality matching.

Example 5.2

Find a maximum cardinality matching in the graph shown below, in which the edges in the current matching are drawn as double lines.

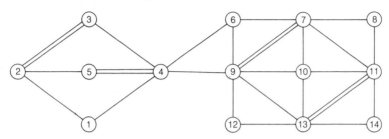

Step 1. The current graph is G_0 with matching M_0. (There are six exposed vertices but there is no edge joining two exposed vertices. We can start building an alternating tree from any one of the six exposed vertices in the graph.)

Step 2. Start an alternating tree T from 1. Then 1 is an outer vertex. Vertex 2 is chosen and it becomes an inner vertex in T along with $\{1, 2\}$. Once 2 is chosen as an inner vertex, its mate 3 is in the tree as an outer vertex along with the matched edge $\{2, 3\}$. At this stage, there are two unexamined edges incident to the current outer vertices: $\{1, 4\}$ and $\{3, 4\}$. Include $\{1, 4\}$ in the tree. So 4 is inner and its mate 5 is outer. The tree at this stage has edges $\{1, 2\}, \{2, 3\}, \{1, 4\}$, $\{4, 5\}$. The only unexamined edge incident to an outer vertex is $\{2, 5\}$, which is incident at 5. If we include $\{2, 5\}$ in the tree, there will be an even cycle. So the edge $\{2, 5\}$ is discarded. There is a Hungarian tree rooted at 1, as shown below.

Step 5. Vertex 1 now becomes examined. Go to step 6.

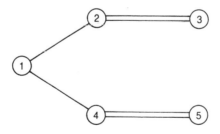

Step 6. There is no blossom to be opened. Go to step 2.

Step 2. Start an alternating tree from 6. Vertex 6 is an outer vertex. Edge $\{6, 7\}$ is marked. Vertex 7 is inner and its mate 9 is outer. Edge $\{6, 9\}$ is marked. This edge joins two outer vertices. So a blossom B_1 is created. The vertices in the blossom are 6, 7 and 9. Shrink the blossom B_1 into a single vertex. The new graph G_1 with matching M_1 is as shown below. (The new graph has more than one exposed vertex.) Return to step 2.

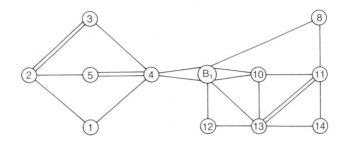

Step 2. Start an alternating tree from the exposed vertex B_1. The edge $\{B_1, 8\}$ is examined. Vertex 8 is not labeled and is exposed. So an augmenting path is obtained. Go to step 3.

Step 3. Add edge $\{B_1, 8\}$ to the matching. The new matching is M_1 as shown below. Go to step 6.

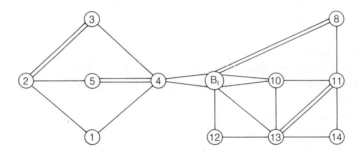

Step 6. Open the blossom B_1. The edge $\{7, 8\}$ is a matched edge. So the edge $\{9, 7\}$ becomes free and $\{6, 9\}$ becomes matched, as below. Go to step 2.

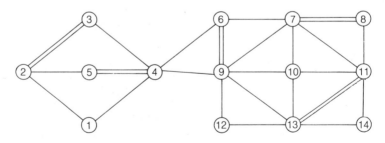

Step 2. Start an alternating tree from the exposed vertex 12. Vertex 12 is outer. Mark edge $\{12, 13\}$. Vertex 13 is exposed but not in the tree, with vertex 11 as mate. So 13 is an inner vertex and 11 is an outer vertex. Mark $\{11, 14\}$. Vertex 14 is not labeled. So an augmenting path has been found from 12 to 14. The new matched edges are $\{12, 13\}$ and $\{11, 14\}$. The edge $\{13, 11\}$ is no longer a matched edge, as shown below. Return to step 2.

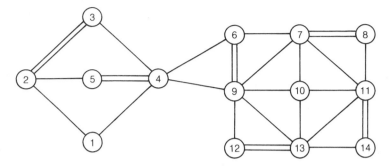

Step 2. There is only one exposed vertex. Go to step 7. (At this stage the only exposed vertex for consideration is vertex 10, since the other exposed vertex (vertex 1) has been already removed for any future consideration.)

Step 7. The current matching (as shown above) is a maximum cardinality matching.

Perfect matchings

A matching M in an undirected graph is a **perfect matching** if every vertex in the graph is incident to an edge in M. If a perfect matching exists for G, then any algorithm to obtain a maximum cardinality matching can be applied to obtain a perfect matching in G. Unlike the cardinality matching algorithm, there is no efficient algorithm to determine whether an arbitrary graph has a perfect matching even though there is a necessary and sufficient condition for the existence of a perfect matching in a graph due to Tuttee (1947). Suppose W is a set of vertices in a graph $G = (V, E)$. Let $\Phi(G - W)$ denote the number of components of $G - W$ with an odd number of vertices. If there is a perfect matching in the graph, then every odd component of $G - W$ has a vertex u which is joined by a matched edge to a vertex in W. So a necessary condition for a perfect matching to exist is that $\Phi(G - W) \leqslant |W|$ for any set W of vertices. If this inequality holds for every set, then it holds for the empty set and so this necessary condition implies that the number of vertices in the graph is even. Tuttee's **perfect matching theorem** (proved in 1947) asserts that this inequality is a sufficient condition for the existence of a perfect matching in an arbitrary undirected graph. An elegant, independent proof of this theorem was given by Lovász (1975), and the reader is referred to that paper for details.

b-matchings

Suppose M is a matching in $G = (V, E)$. Then the degree of each vertex in the subgraph (V, M) is at most equal to 1. This motivates a more general definition of a matching.

Let $V = \{1, 2, \ldots, n\}$. If d_i is the degree in G of vertex i, then the degree vector of G is $\mathbf{d} = [d_1 \ d_2 \ \cdots \ d_n]$. For any subgraph H of G, the degree of i in H is less than or equal to the degree of i in G. Let $\mathbf{b} = [b_1 \ b_2 \ \cdots \ b_n]$ be a fixed vector. A set F of edges in a graph $G = (V, E)$ is called a **b-matching** in G if the degree of vertex i in the subgraph $H = (V, F)$ is less than or equal to b_i. If $b = [1 \ 1 \ \cdots \ 1]$, then a **b**-matching in G is an ordinary matching. A **b**-matching of maximum cardinality in a graph is a maximum cardinality **b**-matching, and a **b**-matching of maximum weight is a maximum weight **b**-matching. Let F be a **b**-matching in $G = (V, E)$. Edges in F are called **b-matched edges**. Other edges are **b-free**. A vertex i is **b-exposed** if the degree of i in $H = (V, F)$ is less than b_i. Otherwise it is a **b-matched vertex**. A **b-alternating path** is a path in which the edges are alternately **b**-matched and **b**-free. A **b**-alternating path in which the first edge and the last edge are **b**-free is called a **b-augmenting path**. In this case we have the following generalization of Theorem 5.2 due to Berge (1973).

Theorem 5.6

A **b**-matching is a maximum cardinality **b**-matching if and only if there is no **b**-augmenting path in the graph. (For a proof of this theorem refer to Gondran and Minoux, 1984.)

5.2 THE STABLE MARRIAGE PROBLEM

Suppose there are n unmarried women and n unmarried men in a community, and that each person is to marry a person of the opposite sex from the community. There are $n!$ ways of arranging a marriage among them. In other words, if each man is represented as a left vertex and each woman is represented as a right vertex and if an edge is drawn joining every left vertex and every right vertex, then the complete bipartite graph has $n!$ complete matchings. If x has to choose between two people y and z, it is possible that x may prefer y to z. In that case it is only fair to assume the edges $e = \{x, y\}$ and $f = \{x, z\}$ are weighted such that $w(e) > w(f)$. So when this issue of preference is introduced, the bipartite graph is indeed a weighted bipartite graph. Even though the bipartite graph is weighted it is possible to obtain an optimization criterion for a 'perfect matching' which does not depend upon the actual weights of the edges. The problem of arbitrary weighted matching will be taken up in section 5.3.

The concept of preference implies that each person in the community ranks members of the opposite sex in accordance with her or his preference for a suitable marriage partner. This preference among possible partners can be represented by an $n \times n$ matrix in which the (i, j)th entry is an ordered pair (p, q), where p is the ranking the right vertex x_i assigns to the left vertex y_j and q is the ranking y_j assigns to x_i. The integers p and q vary from 1 to n.

In this section all matchings are complete matchings. Suppose a matching has already been selected, so everybody in the community now has a spouse.

This matching is said to be an **unstable** matching if there are a man x and a woman y who are not married to each other such that x prefers y to his spouse and y prefers x to her spouse. A matching is **stable** if it is not unstable. Stability is defined in the context of the whole community. It is a global concept.

In a complete matching, suppose the spouses of x_i and x_j are y_i and y_j, respectively. The matching is stable if and only if both the following conditions hold:

1. Either x_i prefers y_i to y_j or y_j prefers x_j to x_i.
2. Either x_j prefers y_j to y_i or y_i prefers x_i to x_j.

The theory of stable marriages presented in this section is based on Gale and Shapley (1962).

Consider the following preference matrix:

$$\begin{bmatrix} (1,3) & (2,2) & (3,1) \\ (3,1) & (1,3) & (2,2) \\ (2,2) & (3,1) & (1,3) \end{bmatrix}$$

in which the rows correspond to three men M_1, M_2 and M_3 and the columns correspond to three women W_1, W_2 and W_3. Under this preference matrix, it is easy to verify that the matching under which each man marries his first choice is a stable marriage even though each woman gets a spouse who is her last choice. It is also easy to see that the mating $(M_1, W_2), (M_2, W_1), (M_3, W_3)$ creates an unstable matching in the community.

It is not at all obvious that for an arbitrary matrix of rankings, there should always be a stable marriage. But this is indeed the case, as shown below. As a matter of fact there is an iterative greedy algorithm which helps us to find a stable matching.

Theorem 5.7
For any preference matrix there exists a stable matching.

Proof
(Since it is almost impossible to improve upon the writing style of Gale and Shapley (1962) the proof is quoted here verbatim. They write 'boy' instead of 'man' and 'girl' instead of 'woman'.)

We shall prove existence by giving an iterative procedure for actually finding a stable set of marriages.

To start, let each boy propose to his favorite girl. Each girl who receives more than one proposal rejects all but her favorite from among those who have proposed to her. However, she does not accept him yet, but keeps him on a string to allow for the possibility that someone better may come along later.

We are now ready for the second stage. Those boys who were rejected now propose to their second choices. Each girl receiving proposals chooses her favorite from the group consisting of the new proposers and the boy on her string, if any. She rejects all the rest and again keeps the favorite in suspense.

We proceed in the same manner. Those who rejected at the second stage propose to their next choices, and the girls again reject all but the best proposal they have had so far.

Eventually (in fact, in at most $n^2 - 2n + 2$ stages) every girl will have received a proposal, for as long as any girl has not been proposed to, there will be rejections and new proposals, but since no boy can propose to the same girl more than once, every girl is sure to get a proposal in due time. As soon as the last girl gets her proposal the 'courtship' is declared over, and each girl is now required to accept the boy on her string.

We assert that this set of marriages is stable. Namely, suppose John and Mary are not married to each other but John prefers Mary to his own wife. Then John must have proposed to Mary at some stage and subsequently been rejected in favor of someone that Mary liked better. It is now clear that Mary must prefer her husband to John and there is no instability.

The condition that the number of men is the same as the number of women can be relaxed.

If there are b boys and g girls with $b < g$, then the procedure terminates as soon as b girls have been proposed to. If $b > g$, the procedure ends when every boy is either on some girl's string or has been rejected by all of the girls. In either case the set of marriages that results is stable.

Gale–Shapley optimality

There is actually a stronger result which assures the existence of an optimal stable matching. Unlike stability, this concept of **optimality** is gender-biased. A stable matching is a **male-optimal stable matching** if every man is at least as well off under it as under any other stable matching. One can define a female-optimal stable matching analogously. (If the set of left vertices represents job applicants and the set of right vertices represents job assignments to different cities, a left-optimal stable assignment makes more sense than a right-optimal stable assignment.) The greedy procedure of Theorem 5.7, in which the men propose to the women, is a male-optimal stable matching. Likewise, if we adopt the greedy procedure where the women propose to the men, then the resulting stable matching is a female-optimal stable matching. There do exist preference matrices in which the male-optimal matching is not the same as the female-optimal matching. The solutions by the two procedures will be the same only when there is a unique stable matching.

The **stable marriage problem** is the problem of finding a stable matching for an arbitrary preference matrix. The constructive proof of Theorem 5.7 gives an iterative algorithm to solve this problem. The matching which results from this 'propose-and-reject' algorithm is not only stable but also optimal.

According to Gale and Shapley (1962) it is possible to define stability in a nonbipartite situation.

> An even number of boys wish to divide up into pairs of roommates. A set of pairings is called stable if under it there are no two boys who are not roommates and who prefer each other to their actual roommates. An easy example shows that there can be situations in which there exists no stable pairing. Namely, consider boys α, β, γ and δ, where α ranks β first, β ranks γ first, γ ranks α first, and α, β and γ all rank δ last. Then regardless of δ's preferences there can be no stable pairing, for whoever has to room with δ will want to move out, and one of the other two will be willing to take him in.

In other words, a stable set (in the Gale–Shapley sense) of homosexual matching need not exist in a community consisting of an even number of people of the same gender.

5.3 WEIGHTED MATCHING PROBLEMS

If $G = (V, E)$ is an arbitrary undirected graph in which each edge e has a nonnegative weight $w(e)$, then the maximum weight matching problem is the problem of finding a matching M such that the sum of the weights of the edges in M is as large as possible. The maximum cardinality matching problem is obviously a special case of this problem when each edge is assigned a weight equal to 1.

Without loss of generality we can assume that the graph is complete because if there is no edge between two vertices u and v, then we can always join them by an edge of weight 0. Furthermore, if the number of vertices is odd, then we introduce one more vertex which is joined to every other vertex by an edge the weight of which is 0. Thus without loss of generality the undirected graph in a weighted matching problem can be considered as a complete graph with an even number of vertices. Finally, we could make one more modification: if w is the maximum among the set of weights of the edges, then we define $w'(e) = w - w(e)$ for each edge in the network. Then the maximum weight matching problem with respect to the weight function w is transformed into a minimum weight matching problem with respect to the weight function w'. Thus the weighted matching problem in this section is the problem of finding a perfect matching in a complete graph (in which each edge has a nonnegative weight) with an even number of vertices such that the sum of the weights of the edges in the matching is as small as possible.

As in section 5.2 we look into bipartite graphs first before embarking upon the more general case. Needless to say, once again it is the existence of odd cycles that makes the nonbipartite matching problem more difficult to handle. The algorithms for solving the weighted matching problems are obtained by an application of the primal–dual algorithm of linear programming, the discussion of which is outside the scope of this book. So the proofs of the results used here will be omitted. Instead, we will explain the procedures for solving these problems and illustrate the procedures by solving two numerical examples. For additional details and an exhaustive treatment of this topic, the reader is referred to Papadimitriou and Steiglitz (1982), Lawler (1976) or Gondran and Minoux (1984). Our discussion here is based on the treatment of this topic by Papadimitriou and Steiglitz.

The weighted bipartite matching problem

In the case of bipartite graphs, the matching problem is known as the assignment problem. In section 2.6 we saw that this problem can be solved by considering the graph as an uncapacitated network and then applying the network simplex method. In the same section another method (the Hungarian method) based on the Konig–Egervary theorem was also discussed. We now discuss a method based on the primal–dual theory of linear programming.

The complete bipartite graph is $H = (V, U; E)$, where both V and U have n vertices. $V = \{v_1, v_2, \ldots, v_n\}$ is the set of left vertices and $U = \{u_1, u_2, \ldots, u_n\}$ is the set of right vertices. The weight of the arc joining v_i and u_j is c_{ij}. The $n \times n$ cost matrix $\mathbf{c} = [c_{ij}]$ need not be symmetric.

A source S and a sink T are introduced as new vertices and arcs are drawn from S to each left vertex and from each right vertex to T. The edge joining v_i and u_j is now considered as an arc from v_i to u_j and is written as (i, j). Thus we have a digraph G (obtained from the bipartite graph H) with $2n + 2$ vertices and $n^2 + 2n$ arcs. The capacity of each arc is 1. There is a feasible flow from the source to the sink in the directed graph G with flow value equal to n.

The central idea behind the algorithm is as follows. At each iteration of the algorithm, we are allowed to send flow from the source to the sink using only some specially chosen arcs (known as admissible arcs) from left vertices to right vertices. So the flow value in the current network (with these admissible arcs) corresponding to the optimal flow may not be equal to n. If the current flow value is n, then we have a minimum weight matching corresponding to the saturated admissible arcs. Otherwise we iterate till the flow value is equal to n. Thus we should know how to construct these admissible arcs at the beginning of each iteration. If the flow value at the current iteration is less than n, then we should also know how to update the flow and move from one iteration to the next iteration with a view to increase the flow value.

At each iteration of the algorithm, we have a vector π consisting of $2n$ components and given by $\pi = [x_1 \ x_2 \ \cdots \ x_n; y_1 \ y_2 \ \cdots \ y_n]$. The set IJ of admissible arcs is the set of arcs (i, j) such that $x_i + y_j = c_{ij}$. Use the arcs in IJ and send as much flow as possible from S to T. This is a maximum flow problem. Let v be the current flow value. If $v = n$, then the admissible arcs define a minimum weight matching and we are done. If $v < n$, we iterate.

For this purpose the vector π has to be updated; this will then define a new set of admissible arcs. Let

$$I^* = \{i: \text{there is a flow augmenting path from } S \text{ to } v_i\}$$
$$J^* = \{j: \text{there is no flow augmenting path from } S \text{ to } u_j\}$$

(A flow augmenting path from s to a vertex is obtained from a directed path from s to that vertex in the associated digraph G' obtained from G using the optimal flow.) Define

$$t = \tfrac{1}{2}\min\{c_{ij} - (x_i + y_j): i \in I^* \text{ and } j \notin J^*\}$$

If $i \in I^*$, then replace x_i by $x_i + t$. Otherwise replace x_i by $x_i - t$. If $j \notin J^*$, then replace y_j by $y_j + t$. Otherwise replace y_j by $y_j - t$. The updated π is used to define the set of admissible arcs for the next iteration.

The algorithm consists of the following steps:

Step 1. Find $\pi = [x_1 \ x_2 \ \cdots \ x_n \ ; y_1 \ y_2 \ \cdots \ y_n]$. (Initially $x_i = 0$ for each i and y_i is the smallest number in column i of the cost matrix).
Step 2. Find the set IJ of admissible arcs.
Step 3. Solve the maximum flow problem using the admissible arcs. If the flow value is n, then go to step 5.
Step 4. Update π. Go to step 2.
Step 5. The set of admissible arcs gives an optimal matching. Stop.

Example 5.3

Obtain a minimum weight matching for the assignment problem with the following 5×5 cost matrix:

$$\mathbf{c} = \begin{bmatrix} 4 & 9 & 3 & 11 & 4 \\ 9 & 8 & 3 & 10 & 8 \\ 7 & 5 & 3 & 8 & 6 \\ 9 & 5 & 3 & 4 & 6 \\ 10 & 11 & 7 & 10 & 11 \end{bmatrix}$$

Iteration 1

Step 1. $\pi = [0 \ 0 \ 0 \ 0 \ 0; 4 \ 5 \ 3 \ 4 \ 4]$.
   Step 2. $IJ = \{(1, 1), (1, 3), (1, 5), (2, 3), (3, 2), (3, 3), (4, 3), (4, 4)\}$.
   Step 3. The optimal flow for the MFP using arcs from IJ is as shown below. The current flow value is 4, which is less than 5.

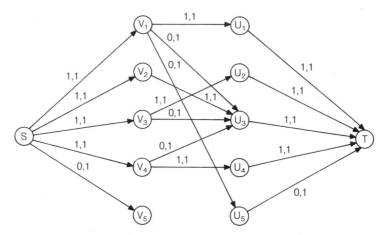

Step 4. There is a flow augmenting path from the source to the left vertex v_5. There are no flow augmenting paths from the source to any of the right vertices.

Thus

$$I^* = \{5\}$$
$$J^* = \{1, 2, 3, 4, 5\}$$
$$t = \tfrac{1}{2}\min\{10 - 4,\ 11 - 5,\ 7 - 3,\ 10 - 4,\ 11 - 4\} = 2$$
$$\pi = [-2\ \ -2\ \ -2\ \ -2\ \ 2;\ 6\ \ 7\ \ 5\ \ 6\ \ 6]$$

Iteration 2

Step 1. $\pi = [-2\ \ -2\ \ -2\ \ -2\ \ 2;\ 6\ \ 7\ \ 5\ \ 6\ \ 6]$.

Step 2. $IJ = \{(1, 1), (1, 3), (1, 5), (2, 3), (3, 2), (4, 3), (4, 4), (5, 3)\}$.

Step 3. Using the current admissible arcs, a flow can be sent saturating the admissible arcs $(1, 5)$, $(2, 3)$, $(3, 2)$, $(4, 4)$ and $(5, 1)$ with flow value 5. So the current solution is optimal.

Step 5. A minimum weight matching consists of the five edges $\{v_1, u_5\}$, $\{v_2, u_3\}$, $\{v_3, u_2\}$, $\{v_4, u_4\}$ and $\{v_5, u_1\}$, with a total weight of $4 + 3 + 5 + 4 + 10 = 26$.

The weighted nonbipartite matching problem

The undirected graph $G = (V, E)$ under consideration now is a complete graph with n vertices where n is even. The weight matrix (cost matrix) **c** is an $n \times n$ symmetric matrix in which each diagonal element is 0. Let $V = \{1, 2, \dots, n\}$. We shall denote the edge joining i and j by (i, j), even though it is an undirected edge. Since the graph is complete, any subset of V of odd cardinality defines an odd cycle. Thus the number of odd cycles in the graph is $N = 2^{n-1} - n$. The set of odd cycles is enumerated as $S_1, S_2, \dots, S_N$. An odd cycle with $2k + 1$ edges is said to be **full** with respect to a matching if there are k matched edges in the cycle.

The (primal–dual) procedure consists of the following steps.

Step 1. The current matching is M. (As in the bipartite case, at the beginning of each iteration of the algorithm we have a vector π which will help us to update the cardinality of M. In the nonbipartite case this vector has $n + N$ components.)

$$\pi = [\mathbf{x}; \mathbf{z}] = [x_1 \ x_2 \ \cdots \ x_n; z_1 \ z_2 \ \cdots \ z_N]$$

The number x_i corresponds to vertex i. The element z_k corresponds to the odd cycle S_k and $z_k \leqslant 0$. (Initially M is empty, x_i is taken as half the smallest nondiagonal element in row i of the cost matrix, and $z_k = 0$ for each k. Since the number of components of z is large, the subvector $\mathbf{z}$ will not be explicitly written down.)

Step 2. Using the vector π, we construct a set J_E of admissible edges and a set J_B of odd cycles as follows:

$$J_E = \left\{ (i, j): x_i + x_j + \sum_{(i,j)\in S_k} z_k = c_{ij} \right\}$$

$$J_B = \{ S_k: z_k < 0 \}$$

(Initially J_B is empty since $z_k = 0$ for every k.)

Step 3. Construct the graph (V, J_E) in which all the currently matched edges are shown. The matching in (V, J_E) is **proper** if all the cycles in J_B are full.

Step 4. Construct G_J by including all edges from J_E and shrinking all sets of odd cardinality from J_B. (G_J is the **admissible graph** corresponding to J_E and J_B.)

Step 5. Find the maximum cardinality matching in G_J starting from the current matching M. Let G_C be the current graph at the conclusion of the cardinality matching algorithm for G_J.

Step 6. Recover the maximum proper matching M of (G, J_E) from the maximum matching from step 5. If $|M| = n/2$, we stop. Otherwise we iterate.

Step 7. The vertices in G_C (from step 5) are called **pseudovertices**. Each pseudovertex is considered as a set. A pseudovertex is an **outer pseudovertex** if it is exposed or if there is an alternating path to it from an exposed vertex in which the last edge is a matched edge. The mate of an outer pseudovertex is an **inner pseudovertex**. (If O' is the set of outer pseudovertices and I' is the set of inner pseudovertices, then the number of exposed pseudovertices in G_C will be $|O'| - |I'|$.)

Let Ψ_O denote the set of pseudovertices (with cardinality at least 3) which are maximal odd sets corresponding to outer pseudovertices or blossoms. Likewise let Ψ_I denote the set of inner pseudovertices (of cardinality at least 3) which are maximal odd sets corresponding to inner pseudovertices.

Step 8. A vertex in V is called an **outer vertex** if v is an element of an outer pseudovertex. It is an **inner vertex** if it is an element of an inner pseudovertex. O is the set of outer vertices and I is the set of inner vertices.

Step 9.

$$\delta_1 = \tfrac{1}{2}\min\{c_{ij} - (x_i + x_j): i \in O, j \in O, i \text{ and } j \text{ do not}$$
$$\text{belong to the same pseudovertex}\}$$
$$\delta_2 = \min\{c_{ij} - (x_i + x_j): i \in O, j \in V - O - I\}$$
$$\delta_3 = \min\{-\tfrac{1}{2}z_k: S_k \in \Psi_I\}$$
$$t = \min\{\delta_1, \delta_2, \delta_3\}$$
$$J_B = \{S_k \in J_B \cup \Psi_O: z_k < 0\}$$

Step 10. (Updating the vector π)

$$x_i = \begin{cases} x_i + t & \text{if } i \in O \\ x_i - t & \text{if } i \in I \\ x_i & \text{otherwise} \end{cases}$$

$$z_k = \begin{cases} z_k - 2t & \text{if } S_k \in \Psi_O \\ z_k + 2t & \text{if } S_k \in \Psi_I \\ z_k & \text{otherwise} \end{cases}$$

Example 5.4

Find a minimum weight matching in the graph the weight matrix of which is

$$\mathbf{c} = \begin{bmatrix} 0 & 6 & 7 & 4 & 3 & 8 & 5 & 7 \\ 6 & 0 & 5 & 6 & 5 & 9 & 8 & 5 \\ 7 & 5 & 0 & 7 & 3 & 2 & 1 & 4 \\ 4 & 6 & 7 & 0 & 4 & 6 & 4 & 5 \\ 3 & 5 & 3 & 4 & 0 & 3 & 7 & 6 \\ 8 & 9 & 2 & 6 & 3 & 0 & 4 & 7 \\ 5 & 8 & 1 & 4 & 7 & 4 & 0 & 6 \\ 7 & 5 & 4 & 5 & 6 & 7 & 6 & 0 \end{bmatrix}$$

The complete graph in this problem is $G = (V, E)$, where $V = \{1, 2, 3, 4, 5, 6, 7, 8\}$. So the number of odd cycles is $2^7 - 8 = 120$.

Iteration 1

Step 1. $\pi = [\mathbf{x}; \mathbf{z}]$ where

$$\mathbf{x} = [\tfrac{3}{2} \ \tfrac{5}{2} \ \tfrac{1}{2} \ \tfrac{4}{2} \ \tfrac{3}{2} \ \tfrac{2}{2} \ \tfrac{1}{2} \ \tfrac{4}{2}]$$

and $\mathbf{z}$ is the zero vector with 120 components.

Step 2. $J_E = \{(1, 5), (3, 7)\}$ and J_B is empty.

Step 3. In the graph (V, J_E), the edges are $(1, 5)$ and $(3, 7)$. There are no matched edges in (V, J_E).

Step 4. There are no odd cycles to be condensed in (V, J_E) and the graph G_J is as in step 3.

Step 5. The maximum cardinality matching in G_J makes the edges $(1, 5)$ and $(3, 7)$ matched edges in the graph G_C.

Step 6. The recovered matching in (V, J_E) is $M = \{(1, 5), (3, 7)\}$. The cardinality of M is less than $n/2 = 4$.

Step 7. $O' = \{\{2\}, \{4\}, \{6\}, \{8\}\}$ and I' is empty. The sets Ψ_O and Ψ_I are empty.

Step 8. $O = \{2, 4, 6, 8\}$ and I is empty.

Step 9

$$\delta_1 = \tfrac{1}{2}\min\{\tfrac{3}{2}, \tfrac{11}{2}, \tfrac{1}{2}, \tfrac{6}{2}, \tfrac{2}{2}, \tfrac{8}{2}\} = \tfrac{1}{4}$$

$$\delta_2 = \min\{2, 2, 1, 5, \tfrac{1}{2} \times \tfrac{9}{2}, \tfrac{1}{2}, \tfrac{3}{2}, \tfrac{11}{2}, \tfrac{1}{2}, \tfrac{1}{2} \times \tfrac{3}{2}, \tfrac{7}{2}, \tfrac{3}{2}, \tfrac{5}{2}, \tfrac{7}{2}\} = \tfrac{1}{2}$$

$$\delta_3 = \infty$$

So $t = \tfrac{1}{4}$. J_B is empty.

Step 10. x_2, x_4, x_6 and x_8 will all be increased by $\tfrac{1}{4}$. Other components of π are unaffected.

Iteration 2

Step 1. $\pi = [\mathbf{x}; \mathbf{z}]$, where

$$\mathbf{x} = [\tfrac{6}{4} \ \tfrac{11}{4} \ \tfrac{2}{4} \ \tfrac{9}{4} \ \tfrac{6}{4} \ \tfrac{5}{4} \ \tfrac{2}{4} \ \tfrac{9}{4}]$$

and $\mathbf{z}$ is the zero vector.

Step 2. $J_E = \{(1, 5), (3, 7), (2, 8)\}$ and J_B is empty.

Step 3. The edges in (V, J_E) are the same edges as in step 2 in which $M = \{(1, 5), (3, 7)\}$.

Step 4. The graph G_J is the same graph (V, J_E) as in step 3 since there are no odd cycles for shrinking.

Step 5. The maximum cardinality matching in G_J gives the matching with edges $(1, 5)$, $(3, 7)$ and $(2, 8)$.

Step 6. The recovered matching M is the same as in step 5. The cardinality of M is less than 4.

Step 7. $O' = \{\{4\}, \{6\}\}$ and I' is empty. Both Ψ_O and Ψ_I are also empty.

Step 8. $O = \{4, 6\}$ and I is empty.

Step 9.

$$\delta_1 = \tfrac{1}{2} \times \tfrac{5}{2} = \tfrac{5}{4}$$

$$\delta_2 = \min\{\tfrac{1}{4}, \tfrac{4}{4}, \tfrac{17}{4}, \tfrac{1}{4}, \tfrac{5}{4}, \tfrac{2}{4}, \tfrac{21}{4}, \tfrac{20}{4}, \tfrac{12}{4}, \tfrac{1}{4}, \tfrac{9}{4}, \tfrac{17}{4}\} = \tfrac{1}{4}$$

$$\delta_3 = \infty$$

Thus $t = \tfrac{1}{4}$. J_B is empty.

Step 10. To update, increase x_4 and x_6 by $\tfrac{1}{4}$. Other components do not change.

Iteration 3

Step 1. $\pi = [\mathbf{x}; \mathbf{z}]$, where

$$\mathbf{x} = [\tfrac{6}{4} \ \tfrac{11}{4} \ \tfrac{2}{4} \ \tfrac{10}{4} \ \tfrac{6}{4} \ \tfrac{6}{4} \ \tfrac{2}{4} \ \tfrac{9}{4}]$$

Step 2. $J_E = \{(1, 4), (1, 5), (2, 8), (3, 6), (3, 7), (4, 5), (5, 6)\}$ and J_B is empty.

Step 3. In (V, J_B) the edges are the same as in step 2. The matched edges are $(1, 5), (3, 7)$ and $(2, 8)$:

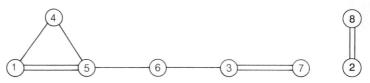

Step 4. The graph G_J is as in step 3, since there are no odd cycles for shrinking.

Step 5. Without destroying the existing matching, obtain a maximum cardinality matching in G_J. The odd set $\{1, 4, 5\}$ is condensed and the graph is as shown below:

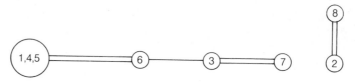

Step 6. The matching is recovered for (V, J_E) as shown below. The cardinality of this matching is 4. Stop.

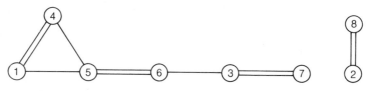

The minimum weight matching is $M = \{(1, 4), (2, 8), (3, 7), (5, 6)\}$ with total weight $4 + 5 + 1 + 3 = 13$.

Example 5.5

Find a minimum weight matching in a graph with weight matrix as given below.

$$\mathbf{c} = \begin{bmatrix} 0 & 2 & 2 & 4 & 5 & 6 \\ 2 & 0 & 2 & 6 & 5 & 4 \\ 2 & 2 & 0 & 7 & 7 & 7 \\ 4 & 6 & 7 & 0 & 4 & 4 \\ 5 & 5 & 7 & 4 & 0 & 4 \\ 6 & 4 & 7 & 4 & 4 & 0 \end{bmatrix}$$

Iteration 1

$\mathbf{x} = [1\ 1\ 1\ 2\ 2\ 2]$ and $\mathbf{z} = 0$. The graph $(V,\ V_E)$ has six edges consisting of the edges belonging to the two odd sets $S_1 = \{1, 2, 3\}$, $S_2 = \{4, 5, 6\}$. No edges are matched. The graph G_J is the same as (V, V_E).

The maximum cardinality algorithm when applied to G_J makes (1, 2) and (4, 5) matched edges. The graph G_C has two condensed vertices S_1 and S_2 with no edges. The recovered matching M has two edges (1, 2) and (4, 5). The cardinality of M is less than 3.

The set O' of outer pseudovertices is $\{S_1,\ S_2\}$ and the set I' of inner pseudovertices is empty.

The set O of outer vertices is $\{1, 2, 3, 4, 5, 6\}$. The set I of inner vertices is empty.

The set Ψ_O of outer pseudovertices is the same as O'.

$\delta_1 = \frac{1}{2}$ and $\delta_2 = \delta_3 = \infty$. So $t = \frac{1}{2}$.

Iteration 2

$\mathbf{x} = [\frac{3}{2}\ \frac{3}{2}\ \frac{3}{2}\ \frac{5}{2}\ \frac{5}{2}\ \frac{5}{2}]$. $S_1 = \{1, 2, 3\}$ corresponds to $z_1 = 0 - 2 \times \frac{1}{2} = -1$. $S_2 = \{3, 4, 5\}$ corresponds to $z_2 = 0 - 2 \times \frac{1}{2} = -1$. $z_k = 0$ for all other k. $J_E = \{(1,\ 4),\ (2,\ 6)\}$. The graph $(G,\ J_E)$ is as follows:

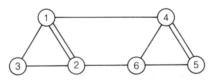

This is the graph G_J:

And this is the graph G_C:

The recovered graph is shown below:

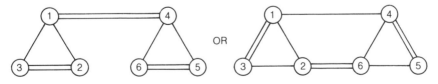

An optimal matching is one of the latter. The weight of minimum matching is $4 + 4 + 2 = 10$.

5.4 THE CHINESE POSTMAN PROBLEM

The Chinese postman problem (CPP) is concerned with situations of the following type: a postman starts from the post office with mail for delivery and delivers it on his beat, covering each street at least once before returning to the post office. It is assumed that while traversing a street in one direction he delivers mail to addresses on both sides of the street by crossing the street back and forth as many times as necessary. In computing the total distance traveled by the postman to complete his delivery, the distance he covers in crossing the streets back and forth is not taken into account. The optimization problem in this context is the problem of finding a closed path starting and ending at the post office such that the length of the path is as small as possible. This problem was posed by Mei-Ko Kwan (1962) and later dubbed the 'Chinese postman problem' by Edmonds (1965b). The highway inspector's problem of finding a route that takes him over each section of the highway at least once is the same as that of the postman. The solution to this problem is also important to repair crews, police patrols and road sweepers.

If the street corners and the post office are considered as vertices, two-way streets joining two street corners as edges and one-way streets as arcs, then the Chinese postman problem can be modeled as a network optimization problem where the weight of an edge (or arc) is the length of the corresponding street. It is reasonable to expect that there are multiple edges and arcs in the network thus constructed. If the graph is undirected and connected, then the problem does have a feasible solution. But in a thoretical sense, if the graph is a directed graph, then the problem need not have a feasible solution. Consider, for example, a street corner which can be reached from the post office but with no streets directed from that corner.

If it is possible for the postman to cover his beat without traversing any edge or arc more than once, then the optimization problem is trivial in the sense that the total distance traveled is the sum of the lengths of all the streets. The problem is thus meaningful only when he has to repeat (deadhead) certain streets to complete his job. In that case the problem is to find a closed path in which the sum of the lengths of the repeated streets is as small as possible.

We distinguish two cases in our study of the CPP: in the first, the graph is undirected; in the second, it is directed.

Undirected graphs

An undirected graph is called an **Eulerian graph** if there exists a closed path in it which uses each edge exactly once. A closed path with this property is called an **Eulerian circuit**. If an Eulerian circuit exists, then it leaves each vertex as many times as it enters and the degree of each vertex is even. The classic theorem in graph theory is the assertion that a graph is Eulerian if and only if it is connected and the degree of each vertex is even. The proof of the sufficiency part of the theorem, which involves the actual construction of an Eulerian circuit, can be found in any elementary text on graph theory.

The construction of an Eulerian circuit in a connected graph G in which the degree of each vertex is even is as follows. We start from any vertex v, and traverse distinct edges until we return to v. This is certainly possible since the degree of each vertex is even. Let C_1 be the circuit (closed path with no repeated edges) thus obtained. If C_1 has all the edges of the graph, then we are done. Otherwise we delete all the edges of the circuit from G and then delete all vertices of degree 0 from the graph. Thus we are left with a subgraph H_1 in which the degree of each vertex is also even. Since G is connected there is a vertex u common to C_1 and H_1. Now we start from u and obtain a circuit C_2 by traversing distinct edges of the subgraph H_1.

Notice that the two circuits have no edges in common but they may have vertices in common. If $u = v$, then we combine the two circuits to obtain a new circuit C_3. If u and v are distinct, then let P and Q be two simple paths between u and v consisting of edges from C_1. Then P, Q and C_2 can be joined together to form a new circuit C_3. If the enlarged circuit C_3 has all the edges of G we are done. Otherwise we continue the process till we obtain an Eulerian circuit.

Example 5.6
Obtain an Eulerian circuit in the graph shown below:

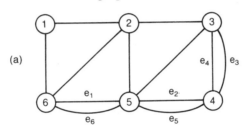

(a)

Starting from vertex 1, we form the circuit C_1 consisting of the edges $\{1, 2\}$, $\{2, 3\}$, e_3, e_2, e_1 and $\{6, 1\}$. Deleting all the edges belonging to C_1 from G and then deleting all vertices of degree 0, we have the subgraph H_1 as shown below.

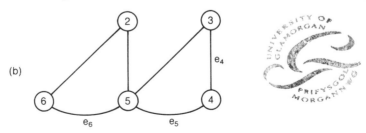

(b)

We see vertex 3 is common to both C_1 and this subgraph. Starting from vertex 3, we get the circuit C_2 consisting of e_4, e_5 and the edge $\{5, 3\}$. Then we combine these two circuits to obtain a larger circuit C_3 which is also not Eulerian. If this circuit as well as all the vertices of degree 0 are deleted, then we have the subgraph H_2 as shown below.

(c)

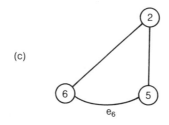

Vertex 2 is common to this subgraph and C_3. We have a circuit C_4 consisting of e_6, $\{2, 5\}$ and $\{6, 2\}$. Finally, we combine C_3 and C_4 to obtain the Eulerian circuit in which the edges are: $\{1, 2\}$, $\{2, 5\}$, e_6, $\{6, 2\}$, $\{2, 3\}$, e_4, e_5, $\{5, 3\}$, e_3, e_2, e_1 and $\{6, 1\}$.

If the graph is not Eulerian, then the postman has to repeat one or more edges. The crux of the problem, then, is to determine the streets to be repeated so that the amount of deadheading is as small as possible.

Thus the problem of solving the CPP in a graph which is not Eulerian is equivalent to the problem of converting the graph into an Eulerian graph by constructing additional edges (by repeating edges) so that the sum of the additional edges is a minimum.

An optimal postman route is obtained by application of the shortest distance algorithm and the minimum weight matching algorithm. Suppose the graph $G = (V, E)$ is not Eulerian. Let V' be the set of vertices of odd degree. The cardinality of V' is even since the sum of the degrees of all the vertices is twice the number of edges. We now construct a complete graph $H = (V', F)$ in which the edge joining the vertices u and v is the shortest distance between u and v in G. We then obtain a minimum weight matching M in H. If e is an edge in M joining two vertices u and v, then we repeat each edge in G of a shortest path in G between u and v. The graph G' obtained by the inclusion of these repeated edges is Eulerian and any Eulerian circuit in G' is a solution of the CPP.

Example 5.7
Find an optimal postman route in the following graph:

(a)

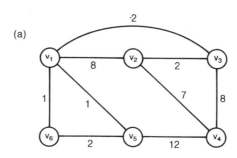

The odd vertices form the set $V' = \{v_2, v_3, v_4, v_5\}$. The complete graph $H = (V'F)$, in which the weight of each edge is the corresponding shortest distance, is as shown below.

(b)

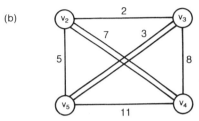

A minimum weight matching in H is $M = \{e, f\}$, where $e = \{v_2, v_4\}$ and $f = \{v_3, v_5\}$. The shortest path between v_2 and v_4 in G is the edge $\{v_2, v_4\}$, which is to be repeated. The shortest path between v_3 and v_5 consists of the edges $\{v_3, v_1\}$ and $\{v_1, v_5\}$ which are also repeated. The enlarged graph, with repeated edges as dashed lines, is as follows:

(c)

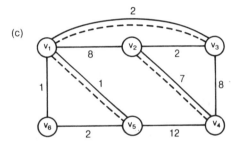

An Eulerian circuit in G' is

$$v_1 - v_2 - v_3 - v_1 - v_3 - v_4 - v_2 - v_4 - v_5 - v_1 - v_5 - v_6 - v_1$$

with a total distance of 53.

Directed graphs

A closed directed path in a digraph is an Eulerian circuit if it contains all the arcs of G. A digraph is Eulerian if it has an Eulerian circuit. If all the streets are one-way streets and if the graph is Eulerian, then a solution of the CPP is obtained by finding any Eulerian circuit in it.

If v is a vertex in a digraph, let $d^+(v)$ denote the outdegree of v and let $d^-(v)$ denote the indegree of v. If a digraph is Eulerian, then it is obvious that G is weakly connected and $d^+(v) = d^-(v)$ for each vertex. As in the case of undirected graphs, it can be shown by a constructive procedure that the converse is also true.

Thus the optimization problem of interest is the following. If G is a digraph which is not Eulerian, then obtain those arcs which have to be repeated so that the enlarged graph is Eulerian and the sum of the weights of the arcs to be repeated is as small as possible.

We conclude this section by showing that the directed CPP can be transformed into an uncapacitated transshipment problem. Let c_{ij} be the weight of the arc (i, j) and x_{ij} be the number of times the arc (i, j) has to be repeated. If there is more than one arc from i to j, then the arc to be repeated is certainly the arc of smallest weight directed from i. So while treating this as a transshipment problem, we choose the arc (i, j) for which c_{ij} is a minimum. The arc (i, j) is chosen accordingly. Then the objective is to minimize the total cost $\Sigma c_{ij} x_{ij}$ under certain constraints. What exactly are those constraints? For one thing, x_{ij} has to be a nonnegative integer for each (i, j). More importantly, at each vertex i the following condition should hold. If the outdegree at a vertex exceeds the indegree by b_i, then the indegree has to be increased by b_i. If the indegree exceeds the outdegree by b_i, then the outdegree has to be increased by b_i. In the transshipment context this means that i is a demand vertex if $b_i = d^+(i) - d^-(i)$ is positive with demand b_i, and i is a supply vertex with supply b_i if b_i is negative. Vertex i is an intermediate vertex if $b_i = 0$. The problem is balanced since the sum of the outdegrees is equal to the sum of the indegrees. The fact that each b_{ij} is a nonnegative integer implies the integrality of the flow vector $\mathbf{x} = [x_{ij}]$.

5.5 EXERCISES

1. An edge in a graph which has a perfect matching is an **unmatchable edge** if no perfect matching in the graph contains that edge. Identify the unmatchable edges in $G = (V, E)$, where $V = \{1, 2, 3, 4, 5, 6\}$ and $E = \{\{1, 2\}, \{1, 6\}, \{2, 3\}, \{2, 4\}, \{2, 5\}, \{2, 6\}, \{3, 4\}, \{4, 5\}, \{5, 6\}\}$.

2. Show that an edge $e = \{u, v\}$ in a graph G with a perfect matching is an unmatchable edge if and only if the graph G' obtained from G by deleting the vertices u and v (and all the edges incident to u and v) has no perfect matching.

3. Show that a tree has at most one perfect matching.

4. In a graph (V, E) with no vertex of degree 0, a set E' of edges is an **edge cover** of the graph if every vertex in the graph is incident to an edge in C. Show that the sum of the number of edges in a smallest edge cover of a graph and the number of edges in a maximum cardinality matching is equal to the number of vertices in the graph G.

5. Given a finite collection of positive integers c_i $(i = 1, 2, \ldots, n)$ which are not necessarily distinct, the problem is to allocate these integers into 'bins' such that the sum of the integers in each bin does not exceed a fixed positive integer B (known as the **bin capacity**) using as few bins as possible. Solve the bin-packing problem (by trial and error) for the particular case where $B = 100$ and the integers are 2, 17, 25, 25, 27, 30, 32 and 40. (There is no known efficient algorithm to solve every instance of this famous optimization problem.)

6. Consider the bin-packing problem with bin capacity B and positive integers c_i, $i = 1, 2, 3, \ldots, n$. Show that if $c_i > B/3$ for each i, then the bin-packing problem can be formulated as a cardinality matching problem.

7. Show that the problem of finding a minimum weight edge cover in a graph can be formulated as a maximum weight **b**-matching problem.

8. Obtain a maximum cardinality matching in the bipartite graph below (the left vertices are X_i, $i = 1, 2, ..., 6$), in which the matched edges in the current matching are drawn as double lines by applying the cardinality matching algorithm.

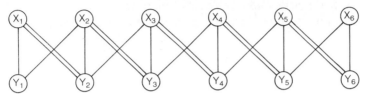

9. Obtain a maximum cardinality matching in the following nonbipartite graph, in which the matched edges under the current matching are marked as double lines.

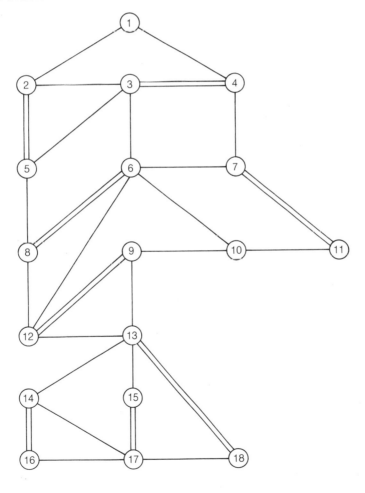

10. Prove that in a male-optimal matching, each woman has the least preferred partner that she can have under any stable matching.

11. Obtain a female-optimal matching for the preference matrix given in section 6.3.

12. Show that a set of marriages is both male-optimal and female-optimal if and only if it is the only stable set of marriages.

13. Obtain a male-optimal matching for the following preference matrix:

$$\begin{bmatrix} (2,3) & (1,3) & (3,1) & (4,1) \\ (1,2) & (2,4) & (3,3) & (4,4) \\ (3,4) & (2,2) & (1,2) & (4,2) \\ (3,1) & (1,1) & (2,4) & (4,3) \end{bmatrix}$$

14. Obtain a female-optimal matching for the preference matrix given in the previous problem.

15. A **b**-matching H with $\mathbf{b} = [b_1 \ b_2 \ \cdots \ b_n]$ in a graph $G = (V, E)$, where $V = \{1, 2, \ldots, n\}$, is a perfect **b**-matching if the degree of vertex i in H is b_i for each i. Construct a graph G' from G as follows. Corresponding to each vertex i in G there will be b_i vertices in G'. Two vertices in G' are joined by an edge e' if and only if these vertices correspond to vertices in G which are endpoints of an edge e in G. If the weight of e in G is $w(e)$, then the weight of e' is G' is also $w(e)$. Show that the problem of finding a perfect **b**-matching of maximum weight in G can be formulated as the problem of finding a maximum weight perfect matching in G.

16. In the graph below, the weight of each edge is marked on the edge. Find a maximum weight perfect **b**-matching if $\mathbf{b} = [2 \ 2 \ 1 \ 1 \ 2]$.

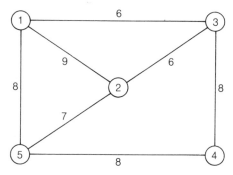

17. By applying the primal–dual method, solve the minimum weight assignment problem represented by the following matrix:

$$\begin{bmatrix} 8 & 9 & 9 & 1 & 12 \\ 3 & 2 & 11 & 1 & 9 \\ 10 & 2 & 10 & 4 & 9 \\ 9 & 6 & 8 & 3 & 1 \\ 12 & 6 & 6 & 6 & 11 \end{bmatrix}$$

18. Find a minimum weight perfect matching in the complete graph equipped with the following weight matrix:

$$\begin{bmatrix} 0 & 12 & 10 & 12 & 10 & 18 & 16 & 10 \\ 12 & 0 & 14 & 8 & 6 & 16 & 10 & 14 \\ 10 & 14 & 0 & 14 & 6 & 4 & 2 & 8 \\ 12 & 8 & 14 & 0 & 8 & 6 & 14 & 12 \\ 10 & 6 & 6 & 8 & 0 & 12 & 8 & 20 \\ 18 & 16 & 4 & 6 & 12 & 0 & 8 & 14 \\ 16 & 10 & 2 & 14 & 8 & 8 & 0 & 12 \\ 10 & 14 & 8 & 12 & 20 & 14 & 12 & 0 \end{bmatrix}$$

19. Obtain an optimal postman route in the following multigraph:

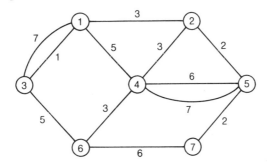

20. Enlarge the undirected network equipped with weight matrix

$$\begin{bmatrix} 0 & 1 & - & - & - & 1 & 4 \\ 1 & 0 & 2 & - & - & - & 1 \\ - & 2 & 0 & 2 & - & - & 4 \\ - & - & 2 & 0 & 3 & - & - \\ - & - & - & 3 & 0 & 9 & 3 \\ 1 & - & - & - & 9 & 0 & - \\ 4 & 1 & 4 & - & 3 & - & 0 \end{bmatrix}$$

by constructing an extra edge joining vertices 1 and 6, with a weight of 7 units, and an extra edge joining vertices 3 and 7, with a weight of 5, so that the resulting network is a weighted multigraph. Obtain an optimal postman route in the enlarged network.

Further reading

Ahuja, R. K., Magnanti, T. L. and Orlin, J. B. (1993) *Network Flows: Theory, Algorithms and Applications*, Prentice-Hall, Englewood Cliffs, NJ.

Lovász, L. and Plummer, M. D. (1986) *Matching Theory*, North-Holland, Amsterdam.

Nemhauser, G. L. and Wolsey, L. A. (1988) *Combinatorial Optimization*, Wiley Interscience, New York.

Schrijver, A. (1986) *Theory of Linear and Integer Programming*, Wiley Interscience, New York.

Syslo, M. M., Deo, N. and Kowalik, J. S. (1983) *Discrete Optimization Algorithms*, Prentice-Hall, Englewood Cliffs, NJ.

References

Bellmore, M. and Malone, J. C. (1971) Pathology of traveling salesman subtour elimination algorithms. *Operations Research*, **19**, 278 307.

Berge, C. (1973) *Graphs and Hypergraphs*, North-Holland, Amsterdam.

Chvátal, V. (1980) *Linear Programming*, W. H. Freeman & Co., San Francisco.

Cunningham, W. H. (1976) A network simplex method. *Mathematical Programming*, **11**, 105–16.

Dantzig, G. B. (1963) *Linear Programming and Extensions*, Princeton University Press, Princeton, NJ.

Dantzig, G. B. and Hoffman, A. J. (1956) Dilworth's theorem on partially ordered sets. *Annals of Mathematics*, **38**, 207–14.

Dijkstra, E. (1959) A note on two problems in connection with graphs. *Numerical Mathematics*, **1**, 269–71.

Edmonds, J. (1965a) Paths, trees and flowers. *Canadian Journal of Mathematics*, **17**, 449–67.

Edmonds, J. (1965b) The Chinese postman problem. *Operational Research*, **13**, Suppl. 1, 373.

Floyd, R. W. (1962) Algorithm 97: Shortest path. *Communications of the ACM*, **5**, 345.

Gale, D. and Shapley, L. D. (1962) College admissions and stability of marriage. *American Mathematical Monthly*, **69**, 9–14.

Gondran, M. and Minoux, M. (1984) *Graphs and Algorithms*, Wiley Interscience, New York.

Hitchcock, F. L. (1941) The distribution of a produce from several sources to numerous localities. *Journal of Mathematical Physics*, **20**, 224–30.

Hoffman, A. J. and Wolfe, P. (1985) The history, in *The Traveling Salesman Problem* (eds E. L. Lawler, J. K. Lenstra, A. H. G. Rinnoy Kan and D. B. Shmoys), Wiley Interscience, New York.

Hu, T. C. (1982) *Combinatorial Algorithms*, Addison-Wesley, Reading, Mass.

Karp, R. M. (1971) A simple derivation of Edmonds' algorithm for optimum branching. *Networks*, **1**, 262–72.

Kwan, M. (1962) Graphic programming using odd and even points. *Chinese Mathematics*, **1**, 273–7.

Lawler, E. L. (1976) *Combinatorial Optimization*, Holt, Rinehart and Winston, New York.

Little, J., Murty, K., Sweeney, D. and Karel, C. (1963) An algorithm for the traveling salesman problem. *Operations Research*, **11**, 979–89.

Lovász, L. (1975) Three short proofs in graph theory. *Journal of Combinatorial Theory*, **B19**, 111–13.

Papadimitriou, C. H. and Steiglitz, L. (1982) *Combinatorial Optimization: Algorithms and Complexity*, Prentice-Hall, Englewood Cliffs, NJ.

Tuttee, W. T. (1947) The factorisation of linear graphs. *Journal of the London Mathematical Society*, **22**, 107–11.

Warshall, S. (1962) A theorem on Boolean matrices. *Journal of the ACM*, **9**, 11–12.

Winter, P. (1987) Steiner problem in networks: a survey. *Networks*, **17**, 129–67.

Solutions to selected problems

CHAPTER 1

6. (i) We selected the edge (6, 7) of weight 1, the edges (6, 9), (4, 7), (4, 5) and (7, 8) of weight 2 each, the edges (5, 2), (9, 10) and (1, 3) of weight 3 each, and the edges (1, 2) and (10, 11) of weight 4 each. The total weight is 26.

    (ii) If we start from vertex 1, then we choose (1, 3). Then we choose the following edges sequentially: (3, 4), (4, 5), (4, 7), (7, 6), (7, 8), (6, 9), (5, 2), (9, 10) and (10, 11). The total weight is 26.

8. (i) The weight of any minimum weight spanning tree in the network is 53.

    (ii) The weight of any maximum weight spanning tree is 120.

    (iii) If we delete the top left-hand vertex of the network, the weight of a minimum weight spanning tree in the graph thus obtained is 52. The two minimum weight edges adjacent to the deleted vertex have weights 1 and 7. Thus $52 + 1 + 7 = 60$ is a lower bound for any optimal Hamiltonian cycle in the network.

9. (iii) Using Prim's algorithm (matrix method), a minimum weight spanning tree of weight 13 consisting of the edges (1, 3), (1, 4), (1, 5), (2, 3) and (4, 6) is easily obtained.

    (iv) If we delete vertex 2, a minimum weight spanning tree of weight 9 is obtained. The two minimum weight edges adjacent to vertex 2 have weights 4 and 4. So a lower bound for an optimal Hamiltonian cycle is $9 + 4 + 4 = 17$.

    (v) Starting from vertex 1, one can obtain the Hamiltonian cycle $1—5—6—4—2—3—1$, with weight 19, by this method.

10. The weight of any spanning tree of minimum weight which contains the edges (1, 2), (1, 3) and (1, 4) is 29.

11. Label the vertices at the level below vertex 1 along the first line as A, B, E, F and H, from left to right. Label the vertices in the next line as D, C,

G and I, also from left to right. A minimum weight arborescence rooted at vertex 1 has the following arcs: $(1, A)$, (A, B), (B, C), (C, D), (C, G), (G, F), (F, E), (F, H) and (H, I), with weight $4 + 2 + 3 + 2 + 4 + 3 + 3 + 4 + 2 = 27$.

12. The vertices have the same labels as in Exercise 11. A maximum weight branching (actually an arborescence) has arcs $(1, A)$, $(1, B)$, $(1, F)$, $(1, H)$, (B, C), (C, D), (C, E), (H, I) and (I, G), with weight $4 + 5 + 6 + 4 + 3 + 2 + 6 + 2 + 5 = 37$.

13. The vertices have the same labels as in Exercise 11. A maximum weight branching (which turns out to be an arborescence) consists of the following arcs: $(1, B)$, (B, C), (C, D), (C, A), (C, E), (E, G), (G, F), (F, H) and (H, I), with weight $5 + 7 + 6 + 9 + 6 + 6 + 9 + 4 + 7 = 59$.

14. The vertices have the same labels as in Exercise 11. A minimum weight arborescence rooted at vertex 1 has the following arcs: $(1, A)$, $(1, B)$, $(1, F)$, $(1, H)$, (B, C), (C, D), (C, E), (C, G) and (H, I), with a total weight of $4 + 5 + 6 + 4 + 7 + 6 + 6 + 4 + 7 = 49$.

CHAPTER 2

2. (i) $\mathbf{y} = [-3 \ -3 \ -1 \ -2 \ 0 \ 0]$.
    (ii) $\mathbf{cx} = 6 + 9 + 8 + 8 + 14 = 45$; $\mathbf{b}^T\mathbf{y} = (-6 \times -3) + (-9 \times -3) = 45$.
    (iii) $\mathbf{x}$ is optimal since no arc is profitable at this stage.

4. The vector $[10 \ 2 \ 0 \ 4 \ 6 \ 2 \ 0 \ 6 \ 0 \ 0 \ 2]^T$ is an FTS with no profitable arcs.

5. Here $\mathbf{x} = [0 \ 8 \ 4 \ 0 \ 16 \ 12 \ 0 \ 6 \ 6 \ 0 \ 0 \ 12]$ is an FTS, with $\mathbf{y} = [3 \ 2 \ 3 \ 1 \ 6 \ 2 \ 1 \ 0]$. The arc $(8, 2)$ is profitable. If it enters the tree, then all the arcs in the cycle created by the entering arc are forward. An updated feasible flow (not an FTS) has the following form:

$$[0 \ (8 + t) \ (4 + t) \ 0 \ (16 + t) \ 12 \ 0 \ 6 \ (6 + t) \ 0 \ t \ 12]$$

The cost of this flow is $-t$, where t could be an arbitrary large positive number.

6. An optimal solution for this extended problem is $[6 \ 0 \ 0 \ 10 \ 8 \ 11 \ 11 \ 7 \ 0 \ 0 \ 0]$. Thus an initial FTS is $\mathbf{x} = [6 \ 0 \ 0 \ 10 \ 8 \ 11 \ 11 \ 7 \ 0]$.

   After three more iterations we obtain for the given problem an optimal solution $\mathbf{x}^T = [0 \ 6 \ 4 \ 0 \ 12 \ 11 \ 5 \ 7 \ 0]$ and the dual solution $\mathbf{y} = [-6 \ -8 \ -5 \ 3 \ -3 \ 10 \ 0]$. The optimal cost is $\mathbf{cx} = \mathbf{yb} = 218$.

11. The set $W = \{3, 4, 7, 8\}$ is an autonomous set in this problem. The complement of W is $W' = \{1, 2, 5, 6\}$. Now we have two problems, one involving W and the other involving W'.

   The optimal solution has flows 7, 6, 10 along $(3, 4)$, $(8, 7)$ and $(7, 6)$ when we use W. It has flows 7, 5, 4 along the arcs $(1, 2)$, $(2, 6)$ and $(5, 1)$ when we use W'. When we combine these two components together we get the following feasible solution:

$$[7\ 5\ 0\ 7\ 0\ 0\ 4\ 0\ 10\ 0\ 0\ 6]$$

The cost is $14 + 15 + 28 + 8 + 20 + 18 = 103$.

12. (i) $\mathbf{x}^T = [4\ 0\ 3\ 6\ 0]$; $\mathbf{z} = \mathbf{cx} = 83$.
  (ii) $[4\ 0\ 6\ 0\ 3\ 7]$.

13. $x_1 = 0$, $x_2 = 0$, $x_3 = 5$, $x_4 = 0$, $x_5 = 2$ and $z = \mathbf{cx} = 7$.

15.

$$A = \begin{bmatrix}
-1 & -1 & -1 & 1 & 0 & 0 & 0 & 0 & 1 & 0 & 0 & 0 & 0 & 0 \\
0 & 0 & 0 & -1 & -1 & -1 & -1 & 0 & 0 & 1 & 0 & 1 & 0 & 0 \\
1 & 0 & 0 & 0 & 1 & 0 & 0 & 0 & 0 & 0 & 1 & 0 & 0 & 0 \\
0 & 1 & 0 & 0 & 0 & 1 & 0 & 1 & 0 & 0 & 0 & 0 & 0 & 0 \\
0 & 0 & 1 & 0 & 0 & 0 & 1 & -1 & 0 & 0 & 0 & 0 & 1 & 0 \\
0 & 0 & 0 & 0 & 0 & 0 & 0 & 0 & -1 & -1 & -1 & 0 & 0 & -1 \\
0 & 0 & 0 & 0 & 0 & 0 & 0 & 0 & 0 & 0 & 0 & -1 & -1 & 1
\end{bmatrix}$$

and $\mathbf{b}^T = [0\ 0\ 6\ 10\ 8\ -9\ -15]$.

16. Here $\mathbf{c} = [3\ 1\ 0\ 0\ 1\ 1\ 4\ 2\ 1\ 0\ 0\ 0]$,

$$A = \begin{bmatrix}
-1 & -1 & 0 & 1 & 0 & 0 & 0 & 0 & 0 & 0 & 0 & 0 \\
1 & 0 & -1 & 0 & 0 & 0 & 0 & 1 & 0 & 0 & 0 & 0 \\
0 & 0 & 1 & -1 & -1 & 0 & 0 & 0 & 0 & 0 & 0 & 0 \\
0 & 1 & 0 & 0 & 0 & -1 & -1 & 0 & 0 & 0 & 0 & 0 \\
0 & 0 & 0 & 0 & 0 & 1 & 0 & -1 & 0 & -1 & 0 & 0 \\
0 & 0 & 0 & 0 & 1 & 0 & 0 & 0 & -1 & 0 & 1 & 0 \\
0 & 0 & 0 & 0 & 0 & 0 & 1 & 0 & 1 & 0 & 0 & -1 \\
0 & 0 & 0 & 0 & 0 & 0 & 0 & 0 & 0 & 1 & -1 & 1
\end{bmatrix}$$

and $\mathbf{b}^T = [-2\ -1\ 2\ -6\ 1\ 0\ 3\ 3]$. For the ILP, the optimal $\mathbf{x}$ is $[1\ 1\ 2\ 0\ 0\ 7\ 0\ 0\ 3]^T$, with cost $\mathbf{cx} = 21$.

18. This becomes a 7×7 transportation problem with supplies $0 + 10$, $0 + 10$, $0 + 10$, $0 + 10$, $2 + 10$, $3 + 10$, $5 + 10$, demands $0 + 10$, $2 + 10$, $3 + 10$, $5 + 10$, $0 + 10$, $0 + 10$, $0 + 10$ and the following cost matrix:

$$\begin{bmatrix}
0 & M & 1 & M & M & M & M \\
1 & 0 & M & M & M & M & M \\
M & 1 & 0 & M & M & M & M \\
M & M & M & 0 & M & M & M \\
1 & 5 & M & M & 0 & M & M \\
11 & M & M & 20 & M & 0 & 5 \\
M & M & M & 5 & 6 & M & 0
\end{bmatrix}$$

The optimal solution is $[5\ 0\ 2\ 2\ 0\ 3\ 0\ 0\ 5\ 0]$, with cost $5 + 2 + 2 + 33 + 25 = 67$.

19. In the unique cycle created by (S_3, D_4), there are two backward arcs with flow values 30 and 35. So the arc (S_2, D_4), with flow value 30, leaves the tree.

21. The optimal flow matrix is

$$\mathbf{x} = \begin{bmatrix} 0 & 0 & 0 & 25 \\ 15 & 0 & 0 & 10 \\ 0 & 20 & 10 & 0 \end{bmatrix}$$

with cost $\mathbf{cx} = 535$.

25. $\mathbf{A}$ is given by $\mathbf{A} = 0.05\mathbf{P}_1 + 0.07\mathbf{P}_2 + 0.08\mathbf{P}_3 + 0.14\mathbf{P}_4 + 0.28\mathbf{P}_5 + 0.38\mathbf{P}_6$ where the permutation matrices $\mathbf{P}_i$ $(i = 1, 2, 3, 4, 5, 6)$ are as follows:

$$\mathbf{P}_1 = \begin{bmatrix} 0 & 1 & 0 & 0 & 0 \\ 1 & 0 & 0 & 0 & 0 \\ 0 & 0 & 0 & 1 & 0 \\ 0 & 0 & 1 & 0 & 0 \\ 0 & 0 & 0 & 0 & 1 \end{bmatrix} \quad \mathbf{P}_2 = \begin{bmatrix} 1 & 0 & 0 & 0 & 0 \\ 0 & 0 & 1 & 0 & 0 \\ 0 & 1 & 0 & 0 & 0 \\ 0 & 0 & 0 & 0 & 1 \\ 0 & 0 & 0 & 1 & 0 \end{bmatrix}$$

$$\mathbf{P}_3 = \begin{bmatrix} 0 & 1 & 0 & 0 & 0 \\ 0 & 0 & 1 & 0 & 0 \\ 0 & 0 & 0 & 0 & 1 \\ 0 & 0 & 0 & 1 & 0 \\ 1 & 0 & 0 & 0 & 0 \end{bmatrix} \quad \mathbf{P}_4 = \begin{bmatrix} 1 & 0 & 0 & 0 & 0 \\ 0 & 0 & 0 & 0 & 1 \\ 0 & 0 & 1 & 0 & 0 \\ 0 & 1 & 0 & 0 & 0 \\ 0 & 0 & 0 & 1 & 0 \end{bmatrix}$$

$$\mathbf{P}_5 = \begin{bmatrix} 0 & 0 & 0 & 0 & 1 \\ 1 & 0 & 0 & 0 & 0 \\ 0 & 1 & 0 & 0 & 0 \\ 0 & 0 & 0 & 1 & 0 \\ 0 & 0 & 1 & 0 & 0 \end{bmatrix} \quad \mathbf{P}_6 = \begin{bmatrix} 0 & 0 & 1 & 0 & 0 \\ 0 & 0 & 0 & 1 & 0 \\ 0 & 1 & 0 & 0 & 0 \\ 1 & 0 & 0 & 0 & 0 \\ 0 & 0 & 0 & 0 & 1 \end{bmatrix}$$

26. The matching $M = \{\{1, 6\}, \{3, 9\}, \{5, 7\}\}$ is a matching of largest size and its cardinality is 3. The covering $\{3, 4, 6\}$ is a covering of smallest size and its cardinality is also 3.

27. The binary matrix $\mathbf{A}$ has five rows corresponding to the five left vertices and four columns corresponding to the right vertices. The element $a_{ij} = 1$ if and only if there is an edge between vertex i and the right vertex which corresponds to j. We have

$$\mathbf{A} = \begin{bmatrix} 1 & 0 & 0 & 0 \\ 1 & 0 & 0 & 0 \\ 0 & 1 & 0 & 1 \\ 1 & 0 & 0 & 0 \\ 0 & 1 & 0 & 1 \end{bmatrix}$$

The 1s in **A** can be covered by three lines: column 1, row 3 and row 5. The (1, 1)th, (3, 2)th and (5, 4)th entries in **A** constitute an independent set of three elements in **A**.

34. There are two optimal solutions with cost = 20:
    (a) (1, 3), (2, 4), (3, 1), (4, 5), (5, 2);
    (b) (1, 3), (2, 5), (3, 1), (4, 4), (5, 2).

35. (1, 1), (2, 2), (3, 4), (4, 3), (5, 5) with $c = 95$.

38. (1, 4), (2, 3), (3, 1), (4, 7), (5, 6), (6, 5), (7, 2), with a total cost of $13 + 16 + 11 + 13 + 12 + 17 + 13 = 95$.

CHAPTER 3

1. $1 \to 2 \to 3 \to 5 \to 7; 3 \to 4; 3 \to 6$.
2. $1 \to 5 \to 2 \to 3 \to 6; 2 \to 4; 2 \to 7$.
3.

$$
\mathbf{D} = \begin{bmatrix}
0 & 1 & 3 & 5 & 5 & 1 & 2 \\
1 & 0 & 2 & 4 & 4 & 2 & 1 \\
3 & 2 & 0 & 2 & 5 & 4 & 3 \\
5 & 4 & 2 & 0 & 3 & 6 & 5 \\
5 & 4 & 5 & 3 & 0 & 6 & 3 \\
1 & 2 & 4 & 6 & 6 & 0 & 3 \\
2 & 1 & 3 & 5 & 3 & 3 & 0
\end{bmatrix}
\quad \text{and} \quad
\mathbf{P} = \begin{bmatrix}
- & 2 & 2 & 2 & 2 & 6 & 2 \\
1 & - & 3 & 3 & 7 & 1 & 7 \\
2 & 2 & - & 4 & 4 & 2 & 2 \\
3 & 3 & 3 & - & 5 & 3 & 3 \\
7 & 7 & 4 & 4 & - & 7 & 7 \\
1 & 1 & 1 & 1 & 1 & - & 1 \\
2 & 2 & 2 & 2 & 5 & 2 & -
\end{bmatrix}
$$

4.

$$
\mathbf{D} = \begin{bmatrix}
0 & 3 & 2 & 2 & 3 & 4 & 2 \\
- & 0 & -1 & -1 & 2 & 1 & 0 \\
- & 1 & 0 & 0 & 3 & 2 & 1 \\
- & 1 & 0 & 0 & 3 & 2 & 1 \\
- & 0 & -1 & -1 & 0 & 1 & -1 \\
- & -1 & -2 & -2 & 1 & 0 & -1 \\
- & 1 & 0 & 0 & 3 & 2 & 0
\end{bmatrix}
$$

and

$$
\mathbf{P} = \begin{bmatrix}
1 & 5 & 5 & 5 & 5 & 5 & 5 \\
1 & 2 & 3 & 4 & 5 & 3 & 7 \\
1 & 6 & 3 & 6 & 5 & 6 & 7 \\
1 & 3 & 3 & 4 & 3 & 3 & 3 \\
1 & 2 & 2 & 2 & 5 & 2 & 7 \\
1 & 2 & 2 & 2 & 2 & 6 & 2 \\
1 & 6 & 6 & 6 & 6 & 6 & 7
\end{bmatrix}
$$

5. $4 \to 3 \to 6 \to 2$ with a shortest distance of 1.

6.
$$1 \rightarrow 5 \rightarrow 7 \rightarrow 6; 5 \rightarrow 2 \rightarrow 3; 2 \rightarrow 4$$

or

$$1 \rightarrow 5 \rightarrow 7 \rightarrow 6 \rightarrow 2 \rightarrow 4 \rightarrow 3$$

7. $2 \rightarrow 4 \rightarrow 3 \rightarrow 6 \rightarrow 2$ is a negative cycle. Another one is $2 \rightarrow 4 \rightarrow 3 \rightarrow 7 \rightarrow 6 \rightarrow 2$.

8. In the iteration, we look at A_4 and P_4 to avoid vertices 5, 6 and 7.

$$A_4 = \begin{bmatrix} 0 & 13 & 4 & 10 & 3 & 6 & 5 \\ - & 0 & -1 & -1 & 2 & 1 & 0 \\ - & 9 & 0 & 8 & 3 & 2 & 1 \\ - & 4 & 0 & 0 & 3 & 2 & 1 \\ - & 0 & -1 & -1 & 0 & 1 & -1 \\ - & -1 & -2 & -2 & 1 & 0 & -1 \\ - & 4 & 3 & 3 & 6 & 2 & 1 \end{bmatrix}$$

and

$$P_4 = \begin{bmatrix} 1 & 3 & 3 & 4 & 5 & 3 & 3 \\ 1 & 2 & 3 & 4 & 5 & 3 & 7 \\ 1 & 2 & 3 & 4 & 5 & 6 & 7 \\ 1 & 2 & 3 & 4 & 3 & 3 & 3 \\ 1 & 2 & 2 & 2 & 5 & 3 & 7 \\ 1 & 2 & 2 & 2 & 2 & 6 & 2 \\ 1 & 2 & 3 & 2 & 2 & 6 & 7 \end{bmatrix}$$

A shortest path from vertex 4 to vertex 2 is $4 \rightarrow 2$.

9.

$$A_3 = \begin{bmatrix} 0 & 20 & 15 & 4 & 3 & 29 \\ 20 & 0 & 19 & 24 & 23 & 9 \\ 15 & 19 & 0 & 8 & 18 & 10 \\ 4 & 24 & 8 & 0 & 6 & 9 \\ 3 & 23 & 18 & 6 & 0 & 7 \\ 29 & 9 & 10 & 9 & 7 & 0 \end{bmatrix} \quad \text{and } P_3 = \begin{bmatrix} 1 & 2 & 3 & 4 & 5 & 2 \\ 1 & 2 & 3 & 1 & 1 & 6 \\ 1 & 2 & 3 & 4 & 1 & 6 \\ 1 & 1 & 3 & 4 & 5 & 6 \\ 1 & 1 & 1 & 4 & 5 & 6 \\ 2 & 2 & 3 & 4 & 5 & 6 \end{bmatrix}$$

A shortest path from 1 to 6 which does not pass through 4 and 5 is $1 \rightarrow 2 \rightarrow 6$ with shortest distance $20 + 9 = 29$.

10. $(3, 6)$ enters and $(2, 6)$ leaves. $1 \rightarrow 3 \rightarrow 6$, with weight $15 + 10 = 25$.

14. $d_i (2, 3, 4, 5, 6, 7, 8) = -, 3, 1, 4, 0, 0, 2$.

15. $A = 1, B = 6, C = 4, D = 7, E = 3, F = 5$ and $G = 2$.

$$A \rightarrow C \rightarrow B; \quad A \rightarrow G \rightarrow E \rightarrow D; \quad E \rightarrow F$$

16. $A = 1, B = 6, C = 2, D = 7, E = 4, F = 8, G = 5$ and $H = 3$. The shortest distances from A to B, C, D, E, F, G and H are respectively $3, -, -6, -,$ $-3, 1$ and $-$.

20. (i) $1\to2\to3\to4\to1$, with weight 32.

    (ii) The revised cost matrix is

$$\begin{bmatrix} - & 5 & 9 & 11 \\ 16 & - & 4 & 7 \\ 21 & 5 & - & 12 \\ 9 & 11 & 6 & - \end{bmatrix}$$

22. The SD matrix is

$$\begin{bmatrix} - & 2 & 3 & 5 & 4 & 7 \\ 3 & - & 1 & 4 & 2 & 6 \\ 2 & 4 & - & 7 & 3 & 5 \\ 4 & 6 & 4 & - & 4 & 2 \\ 6 & 9 & 10 & 2 & - & 4 \\ 4 & 11 & 2 & 4 & 2 & - \end{bmatrix}$$

An optimal cycle is $1\to2\to3\to6\to5\to4\to1$, with weight 12, which is also an optimal salesman circuit.

23. $1\to7\to6\to3\to5\to4\to2\to1$.

24. The vertices have status 17, 14, 19, 25, 26, 22 and 17. Thus vertex 2 is the only median vertex.

25. The only weighted median vertex is vertex 4.

26. Since the cardinality of W is 3, there is at most one Steiner point in the complete graph. By examining the SD matrix the weights of the MST defined by the sets of vertices $\{1, 3, 5\}$, $\{1, 3, 5, 2\}$, $\{1, 3, 5, 4\}$, $\{1, 3, 5, 6\}$ and $\{1, 3, 5, 7\}$ are 8, 7, 8, 9 and 8. So in the complete graph, the Steiner point is at vertex 2. The MST in the complete graph consists of the edges $\{1, 2\}$, $\{3, 2\}$ and $\{5, 2\}$. These shortest paths correspond to the edges $\{1, 2\}$, $\{3, 2\}$, $\{5, 7\}$ and $\{7, 2\}$ which constitute a Steiner tree for W, with Steiner points at vertices 2 and 7.

CHAPTER 4

1. (i) An optimal solution is $[6\ 0\ 4\ 0\ 6\ 8\ 6\ 0\ 0]^T$, with minimum cost $\mathbf{cx} = 126$.

    (ii) An optimal solution is $[4\ 2\ 2\ 0\ 6\ 8\ 6\ 0\ 0]^T$, with minimum cost 138.

2. (i) An optimal solution is $[15\ 0\ 6\ 6\ 0\ 9\ 0\ 0\ 15\ 0]^T$, with minimum cost $\mathbf{cx} = 804$.

    (ii) An optimal solution is the same as in (a) above.

3. The current flow is not optimal. Arc $(1, 3)$ enters and arc $(5, 3)$ leaves, with $t = 8$. The updated solution is $[0\ 8\ 1\ 6\ 4\ 3\ 0\ 0\ 6\ 6\ 8]$, with cost $\mathbf{cx} = 116$. The cost came down by $156 - 116 = 40$.

4. Any arc not in the tree is free for the current solution. The **y**-vector is $[0 \ -16 \ 104 \ 36 \ 56 \ -88 \ -62]$, and we see that $y_j - y_i \leqslant c_{ij}$ for arcs (i, j) not in the tree. So the current solution is optimal.

5. An FTS is $x_{15} = 6$, $x_{23} = 8$, $x_{35} = 18$, $x_{54} = 0$, $x_{56} = 0$, $x_{57} = 10$, $x_{59} = 14$ and $x_{98} = 8$.

6. We obtain an optimal solution for the enlarged graph with positive flow 7 along the artificial arc $(5, 6)$. Thus the given problem is not feasible.

7. The set $W = \{2, 4, 5, 7\}$ is an autonomous set.

8. The set $\{1, 3, 4\}$ is an autonomous set of vertices. The supply at vertex 3 is $3 + 2$ and the demand at vertex 1 is $13 - 2$. At the same time, the demand at vertex 5 is $5 + 2$ and the supply at vertex 2 is $12 - 2$.

9. (i) The feasible flow has positive components $x_{43} = x_{32} = 5$, $x_{21} = 7$, $x_{61} = 3$, with cost $c = 5 + 5 + 7 + 3 = 20$.

   (ii) $W = \{1, 2, 6\}$ is autonomous because the net demand at W is 5, which is equal to its import capacity.

   (iii) The feasible flow in the enlarged network has positive components as in (i) above, and the flow along the artificial arc $(6, 5)$ is 0.

   (iv) The vertices 1, 2, 6 constitute a network, with 2 and 6 as supply vertices with supplies of 7 and 3 units, and 1 as a demand vertex with demand 10 units. The optimal cost for this subproblem is $7 + 3 = 10$. The remaining vertices constitute another network, with 4 as supply vertex with supply 5 units, and 3 as demand vertex with demand 5 units. The optimal cost for this subproblem is 5. Now we have to consider the decomposition cost, which is the import cost using the arc $(3, 2)$. This cost is 5. So the total cost is $10 + 5 + 5 = 20$.

10. Consider the network $G = (V, E)$, with $V = \{1, 2, 3, 4\}$, $\mathbf{x}^T = [x_{12} \ x_{23} \ x_{24} \ x_{34} \ x_{51} \ x_{53}]$ and $\mathbf{b}^T = [-10 \ -3 \ -2 \ 8 \ 7]$. This problem is infeasible. Let the capacity of the arc $(1, 2)$ be 3 and the capacity of every other arc be infinite. In this problem the set $\{2, 3, 4\}$ is autonomous.

11. The maximum flow value is 36 and the minimum cut is $C(S, T)$, where $S = \{1, 2, 3, 4\}$ and $T = \{5, 6\}$.

12. The maximum flow value is 42 and the minimum cut is $C(S, T)$, where $S = \{1, 2, 3, 4, 5, 6, 7\}$ and $T = \{8\}$.

13. The maximum flow value is 32 and the minimum cut is $C(S, T)$, where $S = \{1, 2, 3, 5, 6, 7\}$ and $T = \{4, 8\}$.

14. The maximum flow value is 27 and the minimum cut is $C(S, T)$, where $S = \{1, 3\}$ and $T = \{2, 4, 5, 6, 7\}$.

15. The maximum flow value is 12 and the minimum cut is $C(S, T)$, where $S = \{1, 2\}$ and $T = V - S$.

16. The maximum flow value is 124 and the minimum cut is $C(S, T)$, where $S = \{1, 2, 3, 4, 5, 6, 7, 8\}$ and $T = V - S$.

17. The maximum flow value is 50 and the minimum cut is $C(S, T)$, where $S = \{1, 2, 3, 6, 7, 10\}$ and $T = V - S$.

18. The maximum flow value is 38 and the minimum cut is $C(S, T)$, where $S = \{1, 2\}$ and $T = V - S$.
19. The maximum flow value is 64 and the minimum cut is $C(S, T)$, where $T = \{10, 12\}$ and $S = V - T$.

CHAPTER 5

1. $\{2, 4\}$ and $\{2, 6\}$.
5. At least two bins are needed. Put 17, 25, 27 and 30 in one bin and the others in the other bin. Or put 17, 25, 25 and 32 in one bin and the others in the other bin.
6. Construct a graph with n vertices v_i. Assign the weight c_i to each i. Join v_i and v_j by an edge if and only if $c_i + c_j \leqslant B$. The number of bins needed will be equal to the number of edges in a maximum cardinality matching plus the number of isolated vertices in the graph.
7. Define $b_i = d_i - 1$ for each i. Every edge cover is the complement of a **b**-matching. In particular, if the weight of each edge is 1, then a cover of minimum cardinality can be directly obtained from a maximum cardinality matching.
16. The maximum weight perfect **b**-matching consists of the edges $\{1, 2\}$, $\{2, 5\}$, $\{1, 8\}$ and $\{3, 4\}$, with a total weight of 32.
20. 1—6—1—7—2—1—6—5—7—5—4—3—7—3—2—1, with a total weight of 47.

Index